D0146302

FIXING FREGE

PRINCETON MONOGRAPHS IN PHILOSOPHY

Harry Frankfurt, Editor

•PMP•

The Princeton Monographs in Philosophy series offers short historical and systematic studies on a wide variety of philosophical topics.

Justice Is Conflict by STUART HAMPSHIRE

Liberty Worth the Name by GIDEON YAFFE

Self-Deception Unmasked by ALFRED R. MELE

Public Goods, Private Goods by RAYMOND GEUSS

Welfare and Rational Care by STEPHEN DARWALL

A Defense of Hume on Miracles by ROBERT J. FOGELIN

Kierkegaard's Concept of Despair by MICHAEL THEUNISSEN (translated by Barbara Harshav and Helmut Illbruck)

Physicalism, or Something Near Enough by JAEGWON KIM

Philosophical Myths of the Fall by STEPHEN MULHALL

Fixing Frege by JOHN BURGESS

FIXING FREGE

John P. Burgess

PRINCETON UNIVERSITY PRESS
PRINCETON AND OXFORD

Published by Princeton University Press,
41 William Street, Princeton, New Jersey 08540
In the United Kingdom: Princeton University Press,
3 Market Place, Woodstock,
Oxfordshire OX20 1SY

Library of Congress Cataloging-in-Publication Data

Burgess, John P., 1948–
Fixing Frege / John P. Burgess.
p. cm.—(Princeton monographs in philosophy)
Includes bibliographical references and index.
ISBN-13: 978-0-691-12231-1 (alk. paper)
ISBN-10: 0-691-12231-8 (alk. paper)
1. Logic, Symbolic and mathematical. 2. Frege, Gottlob,
1848–1925—Contributions in mathematics. I. Title. II. Series.

QA9.B854 2005
511.3–dc22 2004054934

British Library Cataloging-in-Publication Data is available

This book has been composed in Janson and Centaur
Printed on acid-free paper. ∞
pup.princeton.edu
Printed in the United States of America
1 3 5 7 9 10 8 6 4 2

For Alexi and Fokion

Contents

Acknowledgments

For so short a work as the present one, the list of persons to whom the author has the pleasant duty of acknowledging debts is a long one: Paul Benacerraf, Solomon Feferman, Kit Fine, Harvey Friedman, A. P. Hazen, Richard Heck, Akihiro Kanamori, Saul Kripke, Stephan Leuenberger, Øystein Linnebo, Marco Lopez, Penelope Maddy, Edward Nelson, Charles Parsons, Stewart Shaprio, Neil Tennant, Kai Wehmeier, and Crispin Wright. Doubtless there have been others whose names have escaped my highly fallible memory, to whom I must offer apologies as well as thanks. Some of those named offered information and assistance on specific points, of which particulars will be noted in the body of the monograph. Others have exerted a global intellectual influence whose contribution to the work is not the less important for being harder to localize. Yet others provided one or another forum in which I could present versions of some of the material for critical comment, and to this last category I must now add Harry Frankfurt, whom I mention in gratitude for the invitation to contribute to the series in which this book now appears. It was a pleasure to work with Ian Malcolm and Sophia Efthimiatou of Princeton University Press, and special thanks are due to Jodi Beder for meticulous care in copyediting.

FIXING FREGE

PMP

I

Frege, Russell, and After

THE GREAT logician Gottlob Frege wrote three books, each representing a stage in a grand program for providing a logical foundation for arithmetic and higher mathematics. His *Begriffsschrift* (1879) introduced a comprehensive system of symbolic logic. The first half of his *Grundlagen* (1884) offered a devastating critique of previous accounts of the foundations of arithmetic, while the second half offered an outline of his own ingenious proposed foundation. The two volumes of the *Grundgesetze* (1893, 1903) filled in the technical details of his outline using his logical symbolism, and extended the project from arithmetic, the theory of the natural numbers, to mathematical analysis, the theory of the real numbers. Unfortunately, just as the second volume of the *Grundgesetze* was going to press Bertrand Russell discovered a contradiction in Frege's system.

Subsequent work in logic and foundations of mathematics largely bypassed the *Grundgesetze* until a couple of decades ago, when philosophers and logicians took a new look at Frege's inconsistent system, and recognized that more can be salvaged from it than had previously been thought. In these last years amended and paradox-free versions of Frege's system have been produced; substantial portions of classical mathematics have been developed within such systems; and a number of workers have claimed philosophical benefits for such an approach to the foundations of mathematics.

The thought underlying the present monograph is that however wonderful the philosophical benefits of Frege-inspired reconstructions of mathematics, the assessment of the ultimate significance of any such approach must await a determination of just how *much* of mathematics can be reconstructed, without resort to *ad hoc* hypotheses, on that approach. What is undertaken in the pages to follow is accordingly a survey of various modified Fregean systems, attempting to determine the scope and limits of each. The present work, though entirely independent of Burgess and Rosen (1997), is thus in a sense a companion to the survey of various nominalist strategies in the middle portions of that work. As in that earlier survey, so in the present one, familiarity with intermediate-level logic is assumed. Boolos, Burgess, and Jeffrey (2002) contains more than enough background material, but neither that nor any other specific textbook is presupposed.

Every strategy, if it is to be consistent, must involve some degree of departure from Frege; but some of the approaches to be surveyed here stay much closer to Frege's own strategy than do others. It is sometimes suggested that the closer one stays to Frege, the greater the philosophical benefits. It is not my aim in the present work to argue for or against such claims. What I do insist is that any philosophical gains must be weighed against mathematical losses. For the survey to follow shows that some approaches yield much more of mathematics than others, and it often seems that the less one keeps of Frege, the more one gets of mathematics. Nonetheless, even in the last system to be considered here, which yields all of orthodox mathematics and more also, there remains one small but significant ingredient of Fregean ancestry.

As a necessary preliminary to the survey of attempts to repair Frege's system, that system must itself be reviewed. The underlying logic of the *Begriffsschrift*, the assumption added thereto in the *Grundgesetze* for purposes of developing mathematics, that development itself, the paradox Russell found in the system, and Russell's own attempts to repair it, must each be briefly examined, and the mathematical and philosophical goals a modified Fregean project, or for that matter any present-day foundational program, might set itself must be briefly surveyed.

1.1 Frege's Logic

While Frege is honored as a founder of modern logic, his system will not look at all familiar to present-day students of the subject. To begin with, Frege uses a non-linear notation that no subsequent writer has found it convenient to adopt, and that will not be encountered in any modern textbook. Since the present work is anything but an historical treatise, the notation will be ruthlessly modernized when Frege's system is expounded here.

But even when the notation is modernized, Frege's *higher-order* logic has a grammar that, though still simpler by far than the grammar of German or English or any natural language, is appreciably more complex than the grammar of the *first-order* logic of present-day textbooks. Nonetheless, after Frege's unfamiliar underlying grammatical and ontological assumptions have been expounded, only a few further explanations should be required to enable the reader familiar with first-order logic to understand higher-order logic.

Let us begin, then, with Frege's grammar. For Frege there are two grammatical categories or grammatical types of what he calls *saturated* expressions. The first, here to be called N, or the category of *names*, includes proper names such as "Plato," but also singular definite descriptions such as "the most famous student of Socrates" that are free from indexicals and designate an *object* (which may be a person or place rather than a "thing" in a colloquial sense), independently of context. The second, here to be called S, or the category of *sentences*, includes declarative sentences such as "Plato was the most famous student of Socrates" that are free from indexicals and have a *truth-value*, true or false, independently of context.

In addition there are many types of *unsaturated* expressions, with one or more gaps that, if filled in with an expression or expressions of appropriate type(s), would produce an expression of type N or S. Those that, thus filled in, would produce expressions of type S rather than type N will be of most interest here, and these may be called *predicates* in a broad sense. In a notation derived from the much later writers Kasimierz Ajdukiewicz and Yehoshua Bar-Hillel, an expression with *k* blanks in it, that if filled in with expressions of types $T_1 \ldots, T_k$ will produce an expression of type S, may be said to be a predicate of type $S/T_1 \ldots T_k$. The simplest case is that of *predicates* in the narrow sense of expressions of type S/N. Some other cases are shown in table A at the back of the book.

The label *relational predicate* will be used for the two-place, three-place, and many-place types S/NN, S/NNN, and so on, with the number of places being mentioned explicitly when it is greater than two. The label *higher predicate* will be used for the second-level, third-level, and higher-level types S/(S/N), S/(S/(S/N)), and so on, with the number of the level being mentioned explicitly if it is higher than the second. Similarly with the label *higher relational predicate*, which even with just two places and even at just the second level covers a variety of types, including not only S/(S/N)(S/N) as shown in the table, but also, for instance, S/(S/NN)(S/NN), and such mixed types as S/(S/N)(S/NN) and even S/N(S/N). It is an instructive exercise to look for natural language examples illustrating such possibilities.

In Frege (1892), which after his three books is its author's most famous work, Frege introduced a distinction between the *sense* expressed and the *referent* denoted by an expression. The reader will not go far wrong who thinks of what Frege calls the "sense" of an expression of whatever type as roughly equivalent to what other philosophers would call its "meaning." What the "referent" of an expression is to be understood to be varies from grammatical type to grammatical type.

In the case of a proper name or singular definite description of type N, the referent is the object designated, the thing bearing the name or answering to the description. Clearly two expressions of type N, for instance, "the most famous student of Socrates" and "the most famous teacher of Aristotle," can have different senses even though they have the same referent—in this instance, Plato. Expressions with different senses but the same referent provide different "modes of presentation" of the same object.

The sense of a sentence of type S Frege calls a *thought*, and the reader will not go far wrong who thinks of what Frege calls a "thought" as roughly equivalent to what other philosophers call a "proposition." The referent of a sentence of type S Frege takes to be simply its truth-value. Obviously two sentences of type S, for instance, "Plato is the most famous student of Socrates" and "Plato is the most famous teacher of Aristotle," or "Plato is a featherless biped" and "Plato is a rational animal," can have different senses, though they have the same referent—in these instances, the truth-value *true*.

So much for the referents of complete or saturated expressions. As for the referents of incomplete or unsaturated expressions of types N/... or S/..., they are supposed to be incomplete, like the expressions themselves, containing gaps that when appropriately filled in will produce an object or a truth-value. The referents of expressions of type N/... Frege calls *functions*, and the referents of expressions of type S/..., that is, the referents of predicates, he calls *concepts*. Corresponding to the different grammatical types of predicates are different ontological types of concepts, including concepts of the narrowest, first-level, one-place kind, but also *relational concepts* and *higher concepts*.

For a concept of type S/N, if filling it in with a certain object produces the truth-value *true*, then the object is said to *fall under* the concept. For instance, assuming for the sake of example that Plato may truly be called wise, since the concept denoted by "… is wise," which is to say the concept of being wise, when filled in by the object denoted by "Plato," which is to say the object Plato, produces the truth-value denoted by "Plato is wise", which is to say the truth-value *true*, it follows that the object Plato falls under the concept of being wise. The same terminology is used in connection with other types of predicates: since Socrates taught Plato, the pair Socrates and Plato, in that order, fall under the relational concept of having taught; since Plato is an example of someone who is wise, the concept of being wise falls under the higher concept of being exemplified by Plato.

Two concepts are called *coextensive* if they apply to the same items, or in other words, if whatever falls under either falls under the other. Now the crucial difference between the *referents* of predicates, which is to say concepts, and the *senses* of predicates is that, according to Frege, coextensive concepts are the same. Thus if every featherless biped is a rational animal and vice versa, then though the senses of "… is a featherless biped" and "… is a rational animal" are different, the concept of being a featherless biped and the concept of being a rational animal are the same.

It sounds odd to say so, and the degree of oddity is a measure of the degree of departure of Frege's technical usage of "concept" from the ordinary usage of "concept," which tends to suggest the sense rather than the referent of a predicate. The label "concept" was in fact chosen by Frege well before he recognized the importance of systematically distinguishing sense and reference. By hindsight it seems he might have done well, after recognizing the importance of that distinction, to revise his terminology. I was tempted to substitute "classification" for "concept" in the foregoing short exposition, but have stuck with "concept" because it is still used by most of the writers with whose views I will be concerned.[1]

In Fregean terminology, then, to the grammatical categories of names, sentences, and predicates there correspond the ontological categories of objects, truth-values, and "concepts." The formal language of Frege's higher-order logic is more complex than the formal languages of the first-order logic expounded in present-day textbooks in that it makes provision for predicates of all types, denoting concepts of all types.

So MUCH FOR THE GRAMMAR behind the logic. Turning to logic itself, modern textbooks introduce the student to the notions of a first-order *language*, with the symbol = for identity, and usually other *non-logical* symbols (*n*-place relation symbols, including 0-place ones or sentence symbols, and *n*-place function symbols, including 0-place ones or constants). Also introduced are *rules of formation*, and the notion of a *term* (built up from variables and constants using function symbols), *atomic formula* (obtained by putting terms in the places of relation symbols), and *formula* (built up from atomic formulas using the logical operators ~, &, ∨, →, ↔, ∀, ∃), along with the ancillary notions of *free* and *bound* variables, and *open* and *closed* formulas (with some variables free and with all variables bound, respectively).[2]

Also introduced are certain *rules of deduction*, which may take widely different forms in different books—the different formats for deductions used in different books going by such names as "Hilbert-style" and "Gentzen-style" and "Fitch-style"—but which lead in all books to equivalent notions of what it is for one formula to be *deducible* from others, and hence to equivalent notions of a *theory*, consisting of all the formulas, called *theorems* of the theory, that are deducible from certain specified formulas, called the non-logical *axioms* of the theory.[3]

Familiarity with all these notions must be presupposed here, but certain conventions that are not covered in all textbooks may be briefly reviewed. To begin with, it proves convenient in practice, when writing out formulas, to make certain departures from what in principle ought to be written. In particular, for the sake of conciseness and clarity certain *defined* symbols are added to the *primitive* or official symbols. The simplest case is the introduction of $\neq$ for *distinctness* by the usual definition, as follows:

(1) $$x \neq y \leftrightarrow \sim x = y$$

What it means to say that $\neq$ is "defined" by (1) is that officially $\neq$ isn't part of the notation at all, and that the left side of (1) is to be taken simply as an unofficial abbreviation for the right side of (1). The "definition" (1) is thus not a substantive axiom, but merely an abbreviation for a tautology of the form $p \leftrightarrow p$. The slash notation for negation may also be used with certain non-logical relation symbols when their shape permits.

Another slightly less simple case, often not covered in textbooks, is that of $\exists!$ for *unique existence*, with two of the several equivalent usual definitions being as follows:

(2a) $\exists! x\phi(x) \leftrightarrow \exists x\phi(x)\ \&\ \forall x_1 \forall x_2(\phi(x_1)\ \&\ \phi(x_2) \rightarrow x_1 = x_2)$

(2b) $\exists! x\phi(x) \leftrightarrow \exists x \forall y(\phi(y) \leftrightarrow y = x)$

Here the first and second conjuncts of the conjunction on the right in (2a) are called the *existence* and *uniqueness* clauses, respectively.

Yet another and less simple case may arise when one has assumed as an axiom or deduced as a theorem something of the form $\exists!\phi(x)$. It may then be convenient to "give a name to" this x by introducing a constant c and assuming $\phi(c)$. The assumption $\phi(c)$ is then called an *implicit* definition, and $\exists!\phi(x)$

the *presupposition* of that definition. Any formula $\psi(c)$ can then be regarded indifferently as abbreviating either of the following:

(3a) $\forall x(\forall y(\phi(y) \leftrightarrow y = x) \rightarrow \psi(x))$

(3b) $\exists x(\forall y(\phi(y) \leftrightarrow y = x) \,\&\, \psi(x))$

I say "indifferently" because (3a) can be deduced from (3b) and vice versa using the presupposition $\exists!\phi(x)$. Note that $\phi(c)$ expands into a logical consequence of that presupposition. Note also that if $\psi(c)$ is of the form $\sim\theta(c)$, it is really a matter of indifference whether one first unpacks the abbreviation in $\theta(c)$ and then applies negation, or first applies negation and then unpacks the abbreviation in $\sim\theta(c)$. For the two results, which read as follows:

(4a) $\sim\forall x(\forall y(\phi(y) \leftrightarrow y = x) \rightarrow \theta(x))$

(4b) $\forall x(\forall y(\phi(y) \leftrightarrow y = x) \rightarrow \sim\theta(x))$

are deducible from each other as were (3a) and (3b); and the same holds for other compounds than negation.

Still yet another and even less simple case is the many-place analogue of the abbreviatory convention just discussed. When one has assumed or deduced $\forall y\exists!x\psi(y, x)$, one may "give a name to" this x by introducing a *function symbol* f and assuming $\forall y\psi(y, f(y))$. The unpacking to eliminate f proceeds analogously to the unpacking to eliminate c in the one-place case. And what has just been said about a one-place function symbol f applies also to many-place function symbols.

Bertrand Russell's famous *theory of descriptions* provides a general notation, which I will write as iota ι, that attaches to a variable x and a formula having x as a free variable to form a term that behaves like an n-place function symbol, where n is the number of free variables other than x. In this notation our constant or 0-place function symbol c above would be $\iota x\phi(x)$, while $f(y)$ would be $\iota x\psi(y, x)$. In both, x counts as a bound variable. Contexts containing ι-terms are expanded as indicated above.[4]

Certain abbreviatory conventions apply specifically to the displaying of laws (axioms and theorems) of a theory. These again may be illustrated by the case of identity, where we have the logical law of *indiscernibility of identicals*, which in many textbook presentations is taken as a logical axiom and in others appears as a logical theorem. Indiscernibility may be stated as follows:

(5) $$x = y \rightarrow (\phi(x) \leftrightarrow \phi(y))$$

One of the conventions illustrated by this way of stating the law is that though we allow ourselves to speak in the singular of "the" axiom or theorem of indiscernibility, what one really has here is an axiom *scheme*, meaning a rule to the effect that all formulas of a certain form are to be counted as axioms, or a theorem *scheme*, meaning a result to the effect that all formulas of a certain form are theorems. What is displayed in (5) is not "the" law, but the general form of an *instance* of the scheme, wherein ϕ may be any formula.[5]

Actually, (5) does not yet fully display the general form of an instance of the law, since it is conventional in displaying laws to omit initial universal quantifiers. Thus what an instance really looks like is this:

(5a) $$\forall x \forall y(x = y \rightarrow (\phi(x) \leftrightarrow \phi(y)))$$

And actually, even (5a) still does not yet fully exhibit the general form of an instance of the law, since it is conventional in displaying laws to omit *parameters*, or additional free variables, that may be present. Thus what an instance *really* looks like is this:

(5b) $$\forall u_1 \ldots \forall u_k \forall x \forall y(x = y \rightarrow (\phi(x, u_1, \ldots, u_k) \leftrightarrow \phi(y, u_1, \ldots, u_k)))$$

Like all the other conventions that have been described so far, these conventions about displaying axioms and theorems of first-order logic will apply also to higher-order logic.

THESE CONVENTIONS HAVE BEEN briefly described here, I say, because not all textbooks cover them. There is one further important topic that very few textbooks cover, that of *many-sorted* first-order languages and theories. A many-sorted language is just like an ordinary one, except that there is more than one style of variables. For instance, in a geometrical theory about points and lines, it may be convenient to have one style of variable X, Y, Z, … for points, and another style of variable ξ, υ, ζ, … for lines. Certain obvious changes in the rules of formation and deduction then have to be made.

As to formation rules and formulas, the usual rules say that $u = v$ is an atomic formula for any variables u and v. In many-sorted logic one has a different identity symbol for each sort of variable. So in our geometrical example we would have atomic formulas of the kinds $\mathrm{X} =_{\mathrm{point}} \mathrm{Y}$ and $\xi =_{\mathrm{line}} \upsilon$, and not atomic formulas identifying a point and a line. Similarly, for non-logical symbols there may be restrictions as to which sorts of variables can go into which places. These changes affect only the rules for forming terms and atomic formulas. The rules for forming more complex formulas from simpler ones by logical operations remain unchanged. In particular, ∀ and ∃ may be applied to any sort of variable.

As to deduction rules and deducibility, the usual rules for quantifiers allow—in one format or another—the inference from $\forall u\phi(u)$ to $\phi(v)$ and from $\phi(v)$ to $\exists u\phi(u)$ for any variables u and v. But in our geometrical example we would want to allow only inference from $\forall \mathrm{X}\phi(\mathrm{X})$ to $\phi(\mathrm{Y})$ and from $\forall\xi\psi(\xi)$ to $\psi(\upsilon)$, and not from $\forall \mathrm{X}\phi(\mathrm{X})$ to $\phi(\upsilon)$ or from $\forall\xi\psi(\xi)$ to $\psi(\mathrm{Y})$. Similarly, for ∃ there are restrictions as to which sorts of variables can be substituted where. These changes affect only the rules of deduction involving quantifiers. The rules for ~, &, ∨, →, ↔ remain unchanged.

In principle, a two-sorted theory can always be replaced by a one-sorted theory with a single style of variables x, y, z, ... by introducing a one-place relation symbol P, called a *sortal predicate*, and replacing quantifications $\forall$x(...) and $\forall\xi$(...) by $\forall x(\mathrm{P}x \rightarrow \ldots)$ and $\forall x(\sim \mathrm{P}x \rightarrow \ldots)$, and $\exists$x(...) and $\exists\xi$(...) by $\exists x(\mathrm{P}x \;\&\; \ldots)$ and $\exists x(\sim \mathrm{P}x \;\&\; \ldots)$. For instance, a geometric theory about points X and lines ξ can be reduced to a one-sorted theory about points-or-lines x by introducing a predicate "is a point," and replacing quantifications "for every point x" and "for every line ξ" by "for every x, if x is a point, then" and "for every x, if x is not a point, then." If there is, say, a constant c of sort x in the original language, we need to add explicitly as an axiom, since it is no longer implicit in the notation, that c is a point: Pc. If there is, say, a function symbol † in the original two-sorted language that takes arguments of sort ξ and gives values of sort x, we need to add explicitly as an axiom, since it is no longer implicit in the grammar, that † applied to a line gives a point: $\sim \mathrm{P}x \rightarrow \mathrm{P}\dagger x$. And similarly for function symbols of more places. A similar reduction can be carried out for three-sorted theories, using two sortal predicates P and Q (and speaking of the items of the third sort as the x such that $\sim \mathrm{P}x \;\&\; \sim \mathrm{Q}x$). But retaining the many-sorted formulation is often in practice more convenient and more illuminating.

For the reader comfortable with first-order logic, including its many-sorted variant version just described, higher-order logic may be introduced as simply one special many-sorted first-order theory, with certain distinctive formulas as axioms, in one special many-sorted first-order language, with certain distinctive relation symbols as primitives. From the perspective of first-order logic, these distinctive primitives and axioms are considered non-logical; from the perspective of higher-order logic, they are considered logical. Whether they are or are not "logical" in a philosophically interesting sense is obviously relevant to assessing the philosophical significance of the development of mathematics within a system based on higher-order logic. But this philosophical issue is irrelevant to the technical definitions of higher-order formula and higher-order deducibility, which as already indicated are the same as for many-sorted first-order logic, which in turn are, apart from some obvious restrictions on which sorts of variables can turn up in which positions, the same as for ordinary, textbook first-order logic, with which it is assumed the reader is familiar.

What the distinctive primitives and axioms of higher-order logic are can almost be guessed from the earlier discussion of Fregean ontology. As to primitives, there are variables of various types T, with the corresponding identity symbols $=_{\mathrm{T}}$, giving rise to such atomic formulas as the following:

$$x =_{\mathrm{N}} y \qquad X =_{\mathrm{S/N}} Y \qquad R =_{\mathrm{S/NN}} S \qquad \mathbf{X} =_{\mathrm{S/(S/N)}} \mathbf{Y}$$

Besides these we will want for each type $\mathrm{T} = \mathrm{S/T_1} \ldots \mathrm{T}_k$ a $(k + 1)$-place relation symbol ∇_{T} to express that a given concept of type T has given items of types $\mathrm{T}_1, \ldots , \mathrm{T}_k$ falling under it. These symbols will give rise to such atomic formulas as the following:

$$\nabla_{\mathrm{S/N}} Xx \qquad \nabla_{\mathrm{S/NN}} Rxy \qquad \nabla_{\mathrm{S/(S/N)}} \mathbf{X}X$$

Such is the language of full *higher-order* logic.[6] If we drop everything above second-level concepts, the result is the language of (*polyadic*) *third-order* logic. If we drop everything above first-level concepts, the result is the language of (*polyadic*) *second-order* logic. If we drop all relational concepts of more than two places, the result is the language of *dyadic higher-order* logic. If we drop all relational concepts whatsoever, the result is the language of *monadic higher-order* logic.

This is all that needs to be said about the *official* notion of formula for higher-order logic. Unofficially, $=_{\mathrm{N}}$ is conventionally just written as $=$, while all other $=_{\mathrm{T}}$ are written as $\equiv$. As for ∇, not only are subscripts dropped, but the very symbol itself is not written, so in practice "x falls under X" and "x and y in that order fall under R" and "X falls under $\mathbf{X}$" are written just Xx and Rxy and $\mathbf{X}X$, or sometimes $X(x)$ and $R(x, y)$ and $\mathbf{X}(X)$.

Turning from primitives to axioms, there are just two of these, called *comprehension* and *extensionality*.

(6) $$\exists X \forall x (Xx \leftrightarrow \phi(x))$$

(7) $$X \equiv Y \leftrightarrow \forall z (Xz \leftrightarrow Yz)$$

The usual conventions for stating axioms apply here: really (6) is an axiom scheme, the formulas displayed are to be prefixed with universal quantifiers, and there may be parameters.[7] A further convention is that when a formula is stated as an axiom of monadic second-order logic, unless explicitly indicated otherwise, the analogous polyadic and higher-order formulas are also to be taken as axioms. Thus in addition to (6) and (7), the following are also comprehension and extensionality axioms:

(6a) $\exists R \forall x \forall y (Rxy \leftrightarrow \psi(x, y))$ (6b) $\exists \mathbf{X} \forall X (\mathbf{X}X \leftrightarrow \theta(X))$

(7a) $R \equiv S \leftrightarrow \forall x \forall y (Rxy \leftrightarrow Sxy)$ (7b) $\mathbf{X} \equiv \mathbf{Y} \leftrightarrow \forall Z (\mathbf{X}Z \leftrightarrow \mathbf{Y}Z)$

Some consequences immediately deducible from these axioms may be noted. (7) implies that the X in (6) is unique, and we may give a name to it, calling it «x: $\phi(x)$», read "the concept of being an x such that $\phi(x)$." In Russellian notation «x: $\phi(x)$» = $\iota X \forall x (Xx \leftrightarrow \phi(x))$. We similarly use the notation «x, y: $\psi(x, y)$» and «X: $\theta(X)$».[8]

Extensionality admits of several equivalent formulations. As formulated in (7) above, together with the indiscernibility of identicals it yields the following:

(8) $$\forall z (Xz \leftrightarrow Yz) \rightarrow (\phi(X) \leftrightarrow \phi(Y))$$

On most contemporary approaches, $\equiv$ is not even included among the official primitives, but rather is regarded as an unofficial abbreviation, in which case (7) is not included among the official axioms, but rather is regarded as the definition of $\equiv$. I will henceforth fall in with this practice. When $\equiv$ is thus taken as defined rather than primitive, it is either (8) that is called the axiom of extensionality, or else the following instance of (8), which together with (6b), actually yields the general scheme (8):

$$\forall z(Xz \leftrightarrow Yz) \rightarrow (\mathbf{X}X \leftrightarrow \mathbf{X}Y) \tag{9}$$

For reasons that will be explained later in this chapter, Frege did not need extensionality in *any* formulation as an axiom for the development of mathematics within his system. Extensionality is needed, however, to express fully his notion of what a concept *is*.

Before closing this section, it will be well to illustrate with an example that plays an important role in Frege's attempt to develop mathematics on a purely logical foundation. To begin with, consider three conditions, each of which may or may not hold for a given relational concept R:

Reflexivity	$\forall x Rxx$
Symmetry	$\forall x \forall y(Rxy \rightarrow Ryx)$
Transitivity	$\forall x \forall y \forall z(Rxy \;\&\; Ryz \rightarrow Rxz)$

Comprehension then gives the existence, and extensionality the uniqueness, conditions in each of the following:

$$\exists! S \forall x \forall y(Sxy \leftrightarrow (x = y \vee Rxy))$$
$$\exists! S \forall x \forall y(Sxy \leftrightarrow (Rxy \vee Ryx))$$
$$\exists! S \forall x \forall y(Sxy \leftrightarrow \forall Z((\forall u(Rxu \rightarrow Zu) \;\&\; \forall u \forall v(Zu \;\&\; Ruv \rightarrow Zv)) \rightarrow Zy))$$

We may give S a name in each case, writing ρR or σR or τR for the S in the three cases. It is easily seen that ρR is reflexive and that σR is symmetric. It is less easily seen but also true—this result having been about the first non-trivial theorem of distinctively higher-order logic, proved in the *Begriffsschrift*—that τR is transitive. ρR and σR and τR are called the reflexive and symmetric and transitive *closures* of R, respectively. If Rxy is the intransitive "x is a parent of y," then τR amounts to the transitive "x is an ancestor of y." On account of this example the transitive closure is also called the *ancestral*. The notion of ancestral was especially important to Frege in his attempted development of mathematics within his system.[9]

In his *Grundlagen*, Frege attempted to situate his philosophical position among others by using Kant's threefold division of knowledge into the *analytic*, the *synthetic a priori*, and the *a posteriori*. In these terms, Frege agreed with the conclusion of Leibniz that arithmetic is analytic, while exposing the fallacy in the argument Leibniz offered in an attempt to establish that conclusion. Frege respectfully disagreed with Kant's claim that arithmetic is synthetic *a priori*, while agreeing with the corresponding claim for geometry—a surprisingly old-fashioned position in the era of non-Euclidean geometry. Frege ridiculed Mill's claim that arithmetic is *a posteriori*.

What was at issue in Frege's disagreements with his predecessors and contemporaries was not the classification of *actual* knowledge. Frege was not interested in how young children learn the basic laws of arithmetic, or how our remote ancestors learned them. He did not claim anyone before himself ever *had* proved the basic laws of arithmetic from principles of pure logic plus appropriate definitions of arithmetical terms in logical terms, without appeal to any distinctively arithmetical "intuition." Frege's claim was, rather, that this *can* be done, even if no one before him ever did it, and that because it can be done arithmetic ranks as analytic.

A potential source of doubt is the principle of mathematical induction, which might be, and by some was, cited as an arithmetical law that could only be established by "intuition."[10] According to this principle, if a condition is fulfilled by zero, and is fulfilled by the successor of any natural number fulfilling it, then it is fulfilled by all natural numbers. Frege already at the time of the *Begriffsschrift* had an idea of how the claim that "intuition" is indispensable for mathematical induction could be refuted.

The Fregean strategy would be first to define the notions of zero and successor, and then to define the greater-than relation as the ancestral of the successor relation.[11] We can then *define* the natural numbers as zero together with those objects that are greater than zero. If we call a concept *inductive* if zero falls under it and the successor of every number falling under it falls under it, then the Fregean definition is more or less equivalent to defining natural numbers as those objects that fall under all inductive concepts. Except for the substitution of talk of a number falling under a concept for talk of a condition being fulfilled by a number—and the two ways of speaking are equivalent by the comprehension axiom—this definition makes the principle of mathematical induction true *almost by definition*.

1.2 Frege's Mathematics

When it came to working out the derivation of arithmetic from logic in detail in his *Grundgesetze*, Frege found he needed one further axiom beyond anything in his *Begriffsschrift*. What need to be explained next are: first, what this additional assumption—which Frege numbered as axiom or basic law V—amounted to; second, how Frege proposed to obtain the most basic laws of arithmetic—the so-called *Peano postulates*—

from it; and third, how one might proceed thereafter to build up higher mathematics.

The new assumption in Frege's *Grundgesetze* is that to every concept there may be associated an object, called its *extension*, in such a way that the extensions of two concepts will be identical objects if and only if the two concepts are coextensive. The most straightforward way to represent this assumption would be as follows. Let us write $€xX$ for "x is the extension of X". Then Frege's assumptions can be expressed as follows:

$$\exists x\, €xX \tag{1}$$

$$€xX \,\&\, €yY \rightarrow (x = y \leftrightarrow X \equiv Y) \tag{2}$$

It is almost immediately deducible from (1) and (2) that we have $\exists!x€xX$. Let us give this x, which in Russellian notation would be $\iota x€xX$, the name $‡X$. Then the following is almost immediate, amounting as it does to little more than a restatement of (2) in this new notation:

$$‡X = ‡Y \leftrightarrow X \equiv Y \tag{3}$$

Several alternative but equivalent formulations of Frege's assumptions are possible, and it will be useful when we come to consider modifications of those assumptions to have set some of these out in advance. A first alternative would be to start with ‡ in place of € as primitive, and (3) in place of (1) and (2) as axiomatic. Then $€xX$ could be defined as $x = ‡X$ and (1) and (2) deduced from (3). This procedure, however, has the serious disadvantage that it *hides the crucial existence assumption in the notation, thus making it almost inaccessible to critical examination and possible revision*, since taking a function symbol as primitive tacitly presupposes that we have $\forall X \exists! x(x = ‡X)$.

Frege usually just called extensions of concepts "extensions of concepts," despite the somewhat unwieldy length of this phrase. He did once remark that what logicians and mathematicians have called "classes" and "sets" are such extensions. And he did sometimes for brevity call his extensions "classes." Here they will instead be called "sets," the reason being that unfortunately a number of later writers have used "classes" in a sense much closer to that of Frege's "concepts" than to his "extensions of concepts." As the word "member" usually goes with "class," so the word "element" usually goes with "set," and will also be used here. A series of abbreviations will now enable us to introduce a notation ßy for "y is a set," as well as the usual notations $x \in y$ for "x is an element of y" and $\{x: \phi(x)\}$ for "the set of x such that $\phi(x)$." For ß and $\in$ we have the following equivalent definitions:

(4a) ß$y \leftrightarrow \exists Y (y = ‡Y)$

(4b) ß$y \leftrightarrow \exists Y$ €yY

(5a) $x \in y \leftrightarrow \exists Y (y = ‡Y \;\&\; Yx)$

(5b) $x \in y \leftrightarrow \exists Y ($€$yY \;\&\; Yx)$

Then the following are almost immediate, amounting as they do to little more than a restatement of (1) and (2) in this new notation:

(6) $\exists y($ß$y \;\&\; \forall x(x \in y \leftrightarrow \phi(x)))$

(7) $\forall y \forall z($ß$y \;\&$ ß$z \;\&\; \forall x(x \in y \leftrightarrow x \in z) \rightarrow y = z)$

Here (6) and (7) are called *comprehension* and *extensionality* for sets (not to be confused with the axioms of the same names for concepts).

Alternatively, one could fall in with the practice of most conventional axiomatic set theories today, which start with an elementhood primitive $\in$, and have a sethood primitive ß also, if like Frege they allow objects that are not sets. Starting with such primitives and (6) and (7) as axioms, one can define €yX to abbreviate the conjunction of ßy and $\forall x(x \in y \leftrightarrow Xx)$, and define ‡ in terms of €, and then deduce ß‡X and $\forall x(x \in ‡X \leftrightarrow Xx)$ and

(3). This procedure, however, departs from Frege's *principle of subordination*, that the relationship of an element to a set is derivative from the relationships of an element to a concept under which it falls, and of a concept to the set that is its extension. The violation is especially serious if one ever contemplates restricting comprehension for concepts, since if $\phi(x)$ is a formula that does not determine a concept, one may have by (6) a set that is the extension of a non-concept.

It is almost immediate from (6) and (7) that we have the following:

$$\exists! y(\text{ß}y \mathbin{\&} \forall x(x \in y \leftrightarrow \phi(x)))$$

We may give a name to this y, calling it $\{x\colon \phi(x)\}$. Alternatively and equivalently, we may define $\{x\colon \phi(x)\} = \ddagger\text{«}x\colon \phi(x)\text{»}$. In either case we have the following:

(8) $\text{ß}\{x\colon \phi(x)\}$

(9) $z \in \{x\colon \phi(x)\} \leftrightarrow \phi(z)$

(10) $\{x\colon \phi(x)\} = \{x\colon \psi(x)\} \leftrightarrow \forall x(\phi(x) \leftrightarrow \psi(x))$

Alternatively, one could start with { : } as primitive and (10) as axiomatic, define $\text{ß}y$ as $\exists Y(y = \{x\colon Yx\})$ and $w \in y$ as $\exists Y(y = \{x\colon Yx\} \mathbin{\&} Yw)$, from which (8) and (9) follow on applying comprehension to ϕ, and from this point one could work backwards to (6) and (7) and if desired from there all the way back to (1) and (2). This procedure, however, has the triple disadvantage that it hides the crucial existence assumption in the notation; that it violates the principle of subordination; and that it produces a formalism *to which the principle tools and results of logical theory developed over the century and more since Frege are not directly applicable*, because the kind of theories for which these tools and results were developed make no provision for an operator that attaches with an object variable to a formula, binding the variable and forming a term for an object.[12]

See table B for a summary of the four options for formalizing Fregean set theory (primitive extension-of relation symbol €, primitive extension-of function symbol ‡, primitive set and element relation symbols ß and ∈, primitive variable-binding term-forming operator { : }). A whole series of further definitions may now be made, which are set down for future reference in table C. Given these definitions, various laws of set theory will follow immediately from well-known laws of logic. For instance, the associative and commutative laws for intersection and union follow immediately from the corresponding laws for conjunction and disjunction.

THE INTRODUCTION OF EXTENSIONS offers an opportunity to dispense with certain other kinds of entities. One such opportunity was not noticed until decades after Frege's work, and so was not used by him, though it has been used by some later writers inspired by Frege. Namely, one can introduce what has become the conventional, if somewhat artificial, Weiner-Kuratowski definition of *ordered* pair, and deduce the basic law for such pairs:

$$\langle x, y\rangle = \{\{x\}, \{x, y\}\}$$
$$\langle x_1, y_1\rangle = \langle x_2, y_2\rangle \leftrightarrow x_1 = x_2 \;\&\; y_1 = y_2$$

Then any purpose that might be served by a relational concept *R* can be served by this one-place concept:

$$«z: \exists x \exists y(z = \langle x, y\rangle \;\&\; Rxy)»$$

In fact the whole many-place apparatus can be similarly dispensed with, leaving us with just monadic higher-order logic.

Another such opportunity was partially, though not wholly, exploited by Frege himself. Namely, any purpose that might be served by a higher concept **X** under which various concepts *Y* fall can be served by this first-level concept:

$$«y: \exists Y(y = \ddagger Y \;\&\; \mathbf{X}Y)»$$

Indeed, by such *level-lowering*, the whole higher-level apparatus can be similarly dispensed with, leaving us with just *second-order logic*. For this or other reasons, most recent writers on modified Fregean systems have confined themselves to the second-order level.

There is yet another option made available by the assumption of extensions, not for dispensing with anything in the system as described so far, but for avoiding the need to introduce something further that it might otherwise seem would be wanted; and this third option Frege does exploit crucially in his treatment of number. The background is as follows. A relational concept R will be called an *equivalence* (*relation*) if it is reflexive, symmetric, and transitive. To give Frege's favorite example, being parallel is an equivalence on lines in the plane. We may then want to introduce *abstracts* with respect to the equivalence, these being, intuitively speaking, objects associated with the original objects in such a way that the abstracts associated to two objects will be the same if and only if the two objects are equivalent. In Frege's favorite example, the abstracts would be the directions of the lines.[13]

More formally, what one is doing here is introducing a new predicate $@_E\, xy$ for "the E-abstract of x is y" with the following assumptions:

$$(11)\qquad \exists z\; @_E xz$$

$$(12)\qquad @_E\, x_1 z_1\ \&\ @_E\, x_2 z_2 \rightarrow (z_1 = z_2 \leftrightarrow E x_1 x_2)$$

$$(13)\qquad \S_E x = \S_E\, y \leftrightarrow Exy$$

Here (11) and (12) give $\exists! z\; @_E\, xz$, and their consequence (13) uses the name $\S_E x$ for this z. As an instance of (13), the direction of one line is the same as that of another if and only if the lines are parallel. Exactly the same thing could be done one level up, starting with a higher concept **E** that is an equivalence and introducing abstracts of first-level concepts. The analogues of (11)–(13) would read as follows:

(11a) $\exists y\ @_{\mathbf{E}}Xy$

(12a) $@_{\mathbf{E}}X_1y_1\ \&\ @_{\mathbf{E}}X_2y_2 \rightarrow (y_1 = y_2 \leftrightarrow \mathbf{E}X_1X_2)$

(13a) $\S_{\mathbf{E}}X = \S_{\mathbf{E}}Y \leftrightarrow \mathbf{E}XY$

Comparing (11a)–(13a) with (1)–(3), it is apparent that extensions are a special case: ‡ is $\S_{\equiv}$.

But this special case subsumes the general case. Extensions of first-level concepts can be used to serve any purpose that would be served by abstracts with respect to equivalences on objects. For given an equivalence E, we may associate with any x its set of equivalents $[x]_E = \{y: Exy\}$. Then we will have, as per (10), that $[x]_E = [y]_E$ if and only if Exy. The abstract with respect to an equivalence may be identified with the set of equivalents. The direction of a line may be taken to be the set of lines parallel to it. Similarly, extensions of second-level concepts can be used to serve any purpose that would be served by abstracts with respect to equivalences on first-level concepts, setting $[X]_{\mathbf{E}} = \{Y: \mathbf{E}XY\}$. And in fact, by exploiting level-lowering, extensions of *first-level* concepts can be used to serve any purpose that would be served by abstracts with respect to equivalences on first-level concepts, using the alternate definition $[X]_{\mathbf{E}} = \{‡Y: \mathbf{E}XY\}$. Again we will then have, as per (10), that $[X]_{\mathbf{E}} = [Y]_{\mathbf{E}}$ if and only if $\mathbf{E}XY$. Thus *to assume extensions is tantamount to assuming abstracts for all equivalences.*

Frege's strategy for developing arithmetic has been exhaustively examined in the recent scholarly literature, and we will be returning to some of its details in a later chapter. For the moment a brief outline will suffice, derived from the scholarly literature to be mentioned in that later discussion, but simplified in a way that involves some departure from historical accuracy, beginning with the wholesale replacement of Frege's notation.

Without further preamble, the notions to be defined are the following: *one-to-one correspondence*, *equinumerosity*, *number*, *zero*, *successor*, and *natural number*. We first define what it is for a relational concept R to be one-to-one correspondence between concepts X and Y, written $X \approx_R Y$, as follows:

$$X \approx_R Y \leftrightarrow \quad \forall x(Xx \rightarrow \exists! y(Yy \ \& \ Rxy)) \ \& \ \forall y(Yy \rightarrow \exists! x(Xx \ \& \ Rxy)) \ \& \ \forall x \forall y(Rxy \rightarrow Xx \ \& \ Yy)$$

We then define what it is for X and Y to be *equinumerous*, written $X \approx Y$, if and only if $\exists R(X \approx_R Y)$.

It is not hard to deduce that equinumerosity is an equivalence, and we have seen how abstracts for equivalences can always be introduced. The abstract for this particular equivalence, $\S_{\approx} X$, will be called the *number* of X, written $\#X$. We thus have the following as an instance of (13a) above:

$$\text{(HP)} \qquad \#X = \#Y \leftrightarrow X \approx Y$$

In connection with this way of introducing *number* Frege cites a vaguely anticipatory remark by Hume—the *serious* antecedent is Cantor—with the result that the last item displayed has come to be called *Hume's principle* (hence the label HP). After having used his theory of extensions to get numbers, for the remainder of his deduction of the laws of arithmetic Frege more or less "forgets about" extensions, and works only from HP.

Writing $\|x: \phi(x)\|$ for $\#«x: \phi(x)»$, the number of x such that $\phi(x)$, then essentially following Frege we may define what it is for x to be a number, written Nx, what the number zero is, written of course 0, and what it is for one number y to be the (immediate) successor of another number x, which I will write

$\$xy$. The definitions read as follows:

$$Nx \leftrightarrow \exists X(x = \|y: Xy\|)$$
$$0 = \|x: x \neq x\|$$
$$\$xy \leftrightarrow \exists X \exists w(x = \#X \ \& \ {\sim}Xw \ \& \ y = \|z: Xz \vee z = w\|)$$

Equivalently, y immediately succeeds x if and only if there are concepts X and Y with numbers x and y such that the objects falling under Y are precisely those falling under X plus one additional object w. We can then define what it is for x to be a natural number, written NNx, in the manner indicated earlier, which makes mathematical induction true almost by definition.

The so-called Peano postulates, whose central role in arithmetic was first recognized by Richard Dedekind, consist of the principle of mathematical induction plus a few further assumptions about zero and successor. In the present context, the further assumptions needed are the following half-dozen items:

(PP1)	$NNx \rightarrow \exists y \ \$xy$
(PP2)	$\$xy \ \& \ \$xz \rightarrow y = z$
(PP3)	$NN0$
(PP4)	$NNx \ \& \ \$xy \rightarrow NNy$
(PP5)	${\sim}\exists x \ \$x0$
(PP6)	$\$x_1y \ \& \ \$x_2y \rightarrow x_1 = x_2$

Here (PP1) says every natural number has a successor, and (PP2) that the successor is unique, so that we may legitimately speak of *the* successor x' of x. Then (PP3) says that zero is a natural number, while (PP4) says that the successor of any natural number is a natural number, so that taking x to x' gives a function on natural numbers. Then (PP5) says that zero is not in the range of this successor function, while (PP6) says that the successor function is one-to-one.

All but one of the (PP) are fairly easy exercises. The exception is (PP1), the *successor theorem*. Frege's idea is to prove this by "bootstrapping." We have zero. We can get one, two, three, and so forth, as follows:

$$1 = \|x{:}\ x = 0\|$$
$$2 = \|x{:}\ x = 0 \vee x = 1\|$$
$$3 = \|x{:}\ x = 0 \vee x = 1 \vee x = 2\|$$

But of course, what really needs to be proved is not the series of statements that zero has a successor one, that one has a successor two, that two has a successor three, and so on; what needs to be proved is, rather, the single statement that *every* natural number has a successor. What the examples suggest, of course, is that if NNx, then $y = \|z{:}\ z < x \vee z = x\|$ will fulfill the condition $\$xy$, where as indicated earlier, order $<$ on natural numbers can be defined in terms of \$ and the ancestral. Turning this heuristic idea into a rigorous proof of the successor theorem turns out to be rather a challenge. In the end, Frege met the challenge, and so established the following:

> ***Frege's Theorem.*** The Peano postulates can be deduced in dyadic second-order logic from Hume's principle and suitable definitions of zero, successor, and natural number.

Refinements, and a converse, will appear in the chapters to follow.

THE NEXT PHASE OF THE PROJECT must be to introduce the operations of addition and multiplication and exponentiation and so forth on natural numbers, and prove the associative, commutative, distributive, and other basic arithmetical laws for these operations. Here there are several quite different ways to proceed, but two stand out as especially significant.

One approach makes constant use of HP, defining operations on numbers in terms of operations on concepts of which they are the numbers. Thus to obtain the sum of x and y, take

X and Y such that $x = \#X$ and $y = \#Y$ and X and Y are disjoint in the sense that no object w falls under both X and Y, and take the number $z = \|w{:}\ Xw \vee Yw\|$ as the sum $x + y$. The definition is a formal counterpart of the idea that the sum of x and y is the number of things you have if you take x things and y other things. The definition has two presuppositions, namely, that for given x and y there will always *exist* such disjoint X and Y, and that the number z obtained in the way described is *unique* in the sense that one gets the same z whichever such disjoint X and Y one takes. Once the definition has been legitimated by proving existence and uniqueness as lemmas, the associative and commutative laws for addition will be immediate from those for disjunction.

On this approach the definition of multiplication will be a formal counterpart of the idea that the product of x and y is the number of ordered pairs that can be formed if each is to consist of a first component chosen from among x things and a second component chosen from among y things, while the definition of exponentiation will be a formal counterpart of the idea that x to the power y is the number of ways of assigning each of y things one of x things. Existence and uniqueness lemmas will have to be proved, but once they are proved, arithmetic laws will be immediate from corresponding logical laws. Generalized beyond the natural numbers to the infinite, this approach leads to Cantor's theory of *transfinite cardinals*.

Another approach "forgets about" HP, and works solely from the Peano postulates. Addition is implicitly defined by the following *recursion equations*:

$$(+0)\quad x + 0 = x$$
$$(+')\quad x + y' = (x + y)'$$

This approach is thus a formal counterpart of the idea that the sum $x + y$ is obtained by starting with x and adding one y times. To legitimate the definition, one needs to prove a lemma to the effect that there exists a unique function for which the recursion equations hold, or equivalently that there exists a unique R for which we have the following:

$$Rx0z \leftrightarrow z = x \qquad Rxy'z \leftrightarrow \exists w(Rxyw \,\&\, z = w')$$

Once the definition has been legitimated, the associative law is then proved by mathematical induction, the base or zero case and the induction or successor case being verified as follows:

$$(x + y) + 0 = x + y = x + (y + 0) \tag{14}$$

$$\begin{aligned}(x + y) + z' &= ((x + y) + z)' = x + (y + z))' \\ &= x + (y + z)' = x + (y + z')\end{aligned} \tag{15}$$

In (14), which shows the associative law holds for $z = 0$, $(+0)$ is used. In (15), which shows the associative law holds for z' assuming it holds for z, $(+')$ is used. Induction is similarly used to prove the commutative law.

On this approach, multiplication, exponentiation, and beyond that, superexponentiation and further operations are likewise implicitly defined by recursion equations which give the value of the operation for 0 outright, and give the value of the operation for z' in terms of the value for z, thus:

$$\begin{array}{llll}(\cdot 0) & x \cdot 0 = 0 & (\cdot ') & x \cdot y' = x + x \cdot y \\ (\uparrow 0) & x \uparrow 0 = 1 & (\uparrow ') & x \uparrow y' = x \cdot (x \uparrow y) \\ (\Uparrow 0) & x \Uparrow 0 = 1 & (\Uparrow ') & x \Uparrow y' = x \uparrow (x \Uparrow y)\end{array}$$

Here one has a formal counterpart of the idea of multiplication as repeated addition, exponentiation as repeated multiplication, and so on. The lemmas needed to legitimate the definition of addition and of those of these further operations can actually be combined into a single grand lemma legitimating *definition by recursion*. Then mathematical induction is used to deduce the relevant laws of arithmetic from the recursion equations. For the natural numbers, this approach is essentially Dedekind's.[14] Hence the name attached to the following result:

> ***Dedekind's Theorem.*** The existence of unique operations on natural numbers satisfying the recursion equations for addition, multiplication, exponentiation, and so on, can be deduced in dyadic second-order logic from the Peano postulates.

Generalized beyond the natural numbers to the infinite, this approach leads to Cantor's theory of *transfinite ordinals*.

Frege was aware of Cantor's work, and valued his theory of the transfinite. However, where Cantor developed the cardinal and ordinal approaches in tandem, Frege (who to be sure gave few details about the development of arithmetic after obtaining the Peano postulates) almost certainly intended that the cardinal, rather than the ordinal, approach should be followed. Of course, a major task on either the cardinal or the ordinal approach would be to prove that it is equivalent to the other. (Or more precisely, that the two approaches are equivalent for natural numbers, one of the striking features of Cantor's theory of the transfinite being that the cardinal and ordinal notions diverge there.) On the cardinal approach, equivalence with the ordinal approach is needed to establish anything like an efficient means by which sums and products can be *calculated*. On the ordinal approach, equivalence with the cardinal approach is needed to establish how addition and multiplication can be *applied*.

Once one has the system of natural numbers $\mathcal{N}$ with the usual arithmetic operations and the usual laws for them, the next task is to obtain the systems of integers $\mathcal{Z}$, of rational numbers $\mathcal{Q}$, of real numbers $\mathcal{R}$, of complex numbers $\mathcal{C}$, and if one is really ambitious, of quaternions $\mathcal{H}$. In each case the numbers of a new kind and the operations on them are defined in terms of the numbers of the previous kind and the operations on *them*, and the arithmetic or algebraic laws for the new kind are deduced from those for the previous kind together with these definitions. But it is not needful to enter into any further details of the construction at this stage.

1.3 Russell's Paradox

Frege himself never got as far as publishing the projected third volume of his *Grundgesetze*, which would have completed his treatment of $\mathcal{R}$ and perhaps included a treatment of $\mathcal{C}$. For just as his second volume was about to see print, Frege received a letter from Russell (1902) exposing a contradiction in Frege's theory of extensions of concepts. Our next order of business must be to review Russell's famous paradox as it applies to Frege's system, and Frege's own immediate reaction to the discovery of the contradiction.

Frege's publications did not at once attract the degree of attention they deserved. Bertrand Russell, for one, learned of them only rather late, after he had already independently rediscovered some of the same ideas. His version can be found in Russell (1903), which corresponds to Frege's *Grundlagen*, in the sense that its ideas are presented in an informal manner. There are, as one would expect in a case of independent discovery, some differences between Frege's approach and Russell's. For one thing, Frege originally (that is to say, in the *Grundlagen*) passed up the opportunity to dispense with higher-level concepts, and defined the number two, for instance, so that it is the extension of such a higher-level concept, namely, that under which fall all and only those first-level concepts under which there fall exactly two objects. By contrast, Russell from the beginning identified the number two with the extension of a first-level concept, that under which there fall all and only those objects that are extensions in first-level concepts under which there fall exactly two objects—or in less Fregean and unwieldy terms, with the set of all sets having exactly two elements.

One feature Russell's approach did share with Frege's is that both recognized a universal set. And Russell, who seems to have read Cantor's work with greater attention than did Frege, noticed that the assumption of such a set was in apparent conflict with the most famous of Cantor's theorems, which says that the number of subsets of a given set is always greater than the number of elements of that set. Cantor himself had noticed this conflict, which indeed is sometimes called *Cantor's paradox*, even earlier; but he kept quiet about it.[15]

Given Cantor's definition of the greater-than relation for numbers (which is equivalent to Frege's for natural numbers, but also applies to transfinite cardinals), what has to be proved to establish that the number of elements in a set T is greater than the number of elements in a set S is that there is no one-to-one correspondence between T and any subset S' of S. Cantor's theorem was thus that there can be no one-to-one correspondence between the subsets of S and the elements of any subset S' of S. What Russell noticed was that if V is the *universal* set of all objects, then every subset of V being a set and therefore an object and therefore an element of V, there is an obvious one-to-one correspondence between the subsets of V and the elements of the subset V′ of V consisting of all sets—namely, the one-to-one correspondence in which each set corresponds to itself.

Now Cantor's general proof proceeds by assuming there *is* a one-to-one correspondence between the subsets of S and the elements of some subset S' of S and deriving a contradiction. The contradiction is obtained by considering the special subset S^* of S' consisting of all elements u of S' such that u is not an element of the set to which it corresponds, and then considering the special element s^* of S' corresponding to S^*. The contradiction is that s^* must be an element of S^* if and only if s^* is *not* an element of S^*. In the special case of the universal set, the special set would be the set of all sets that are not elements of themselves. And in this special case we have the contradiction that this set is an element of itself if and only if it is not an element of itself. (The set of all *objects* that are not elements of themselves, or in other words, that *either* are sets that are not elements of themselves *or* are not sets at all, gives essentially the same contradiction.) This contradiction is *Russell's paradox*.

Having heard of Frege's work, Russell hoped that it might contain some solution to the paradox, showing the contradiction to be merely apparent. Having obtained and examined Frege's writings, Russell concluded that the contradiction is real, and arises in Frege's framework just as it arose in his own. And indeed, just look at one of the laws of Frege's set theory:

$$(1) \qquad z \in \{x: \phi(x)\} \leftrightarrow \phi(z)$$

One gets an immediate contradiction on taking $x \notin x$ for $\phi(x)$ and $\{x: x \notin x\}$ for z, thus:

$$\{x: \sim x \in x\} \in \{x: \sim x \in x\} \leftrightarrow \sim \{x: \sim x \in x\} \in \{x: \sim x \in x\}$$

Frege's system is inconsistent.[16]

THE FIRST PHILOSOPHER-LOGICIAN who attempted to repair the *Grundgesetze* after the discovery of the Russell paradox was Frege himself, in an appendix to the second volume, added at the last minute in response to the letter he had just received from Russell about the paradox. In effect, Frege's proposal—*Frege's way out*—involved a minimal modification of his assumptions about extensions, leading to minimally modified notions of set and element that may be called set* and element*. In place of the law (1), we have only the following minimally modified law, which allows just one exception to (1):

$$(2) \qquad z \in^* \{x: \phi(x)\}^* \leftrightarrow z \neq \{x: \phi(x)\}^* \ \& \ \phi(z)$$

And on account of this one exception, (2) is insufficient for Russell's deduction. Russell's initial opinion of Frege's proposal was that it might well work.

It does not. Or at least, it does not if one assumes that there are at least two distinct objects, which one obviously needs to do if one is going to get anywhere at all in developing mathematics. The problem is the following. Russell considered the set of sets that are not elements of themselves, asked whether it has itself as an element, and obtained a contradiction in this way; but he could alternatively have considered the set of singletons that are not elements of the objects of which they are the singletons, and asked whether it has its own singleton as an element, and he would still have gotten a contradiction. It turns out that Frege's quick fix blocks Russell's actual argument, which turns on asking whether a certain set is an element of *itself*—the law (2) implies that *no* set* is an element* of itself—but not the alternative using singletons.

To see this is so, for any object a consider the singleton* defined by $\{a\}^* = \{x\colon x = a\}^*$. Then by (2) there are just two possibilities. In what may be called the *normal* case, $a \neq \{a\}^*$ and a is the one and only element* of $\{a\}^*$; in the abnormal case, $a = \{a\}^*$ has no elements* at all. It follows that in every case distinct objects have distinct singletons*. Having established the properties of singletons*, consider now the set*

$$r = \{y\colon \exists z(y = \{z\}^* \;\&\; y \notin^* z)\}^*$$

and $s = \{r\}^*$. By the properties of singletons*, r is the only z such that $s = \{z\}^*$, and if r has any element*, we are in the normal case and $s \neq r$. It then follows by (2) that we have $s \notin^* r$ if and only if $s \notin^* r$, a contradiction.

It remains to show that r does indeed have at least one element*. To this end consider $\varnothing^* = \{x\colon x \neq x\}^*$ and $V^* = \{x\colon x = x\}^*$. By (1) no object is an element* of $\varnothing^*$ and every object but V^* itself is an element* of V^*. Since we are assuming there are at least two objects, V^* has at least one element*, and so is distinct from $\varnothing^*$. By the properties of singletons* it then follows that $V^* \in^* \{V^*\}^* \neq V^*$, $\{V^*\}^* \in^* \{\{V^*\}^*\}^* \neq \{V\}^*$, and so on. Moreover, since there are now at least three objects, $\varnothing^*$ and V^* and $\{V^*\}^*$, V^* has at least two elements*, and so is distinct from any singleton*, including $\{\{V^*\}^*\}^*$. Therefore $\{\{V^*\}^*\}^*$ is distinct from the only element* V^* of $\{V^*\}^*$, and we may take $z = \{V^*\}^*$ and $y = \{\{V^*\}^*\}^*$ to get an example of a y and z with $y = \{z\}^*$ and $y \notin^* z$. Then (2) implies that $y \notin^* r$ unless $y = r$, in which latter case $z \in^* r$, so that in either case r has at least one element*, as required to complete the proof.

Thus Russell's paradox cannot be avoided by Frege's minimal modification of his system. This was soon realized by Russell, and realized by Frege somewhat later, though a deduction of a contradiction from Frege's modified system only appeared in the published literature about a half-century after the original paradox.[17] Some more substantial modification in Frege's assumptions will be required.

1.4 Russell's Solution

The second philosopher-logician who attempted to repair the *Grundgesetze* after the discovery of the Russell paradox was Russell himself, in his famous *theory of types*. This theory received its first extended presentation in an article by Russell (1908), which amounted to an announcement of the program to be pursued in the joint work of Whitehead and Russell (1910–13). A simplification was proposed by F. P. Ramsey (1925), and what is generally called the *theory of types* by logicians today derives more from Ramsey than directly from Russell. Thus one must distinguish the original, Russellian theory, called the *ramified* theory of types, from the revised version, called the *simple* or *ramseyfied* theory of types. It is the original theory that will be of most direct interest here.

Underlying the theory was a certain analysis of what was responsible for the Russell paradox. It will be recalled that Russell's route to his paradox began with a conflict between an assumption he shared with Frege, that of the existence of a universal set, and a result of Cantor. The contradiction is not as damaging to Cantor's work as it is to Frege's. For though the contradiction does draw unfavorable attention to the fact that Cantor's publications offered no explicit account of which conditions determine sets, Cantor had certainly never assumed that *every* condition $\phi(x)$ does so, and in particular had never assumed that such a condition as $x \notin x$ does. Yet those who had never liked Cantor's theory, those who had always been suspicious of sets in general and infinite ones in particular, felt justified in their suspicions after the discovery of Russell's paradox, and were quick to conclude that it was the assumption of the existence of a new kind of object, a *set* associated with a formula $\phi(x)$, that was to blame for the contradiction.

But is lack of "ontological parsimony" and acceptance of "actual infinity" really to blame? Not obviously. To assume there is a distinctive object associated with each formula $\phi(x)$ is not in itself absurd, since after all there *is* such an object, namely, the formula $\phi(x)$ itself, considered as a linguistic entity. Suppose that we forget about sets and concepts, too, and consider only linguistic expressions, and specifically formulas $\phi(x)$ with one free variable. Instead of asking whether an object belongs to the set $\{x: \phi(x)\}$, and instead of asking whether an object falls under the concept «x: $\phi(x)$», let us ask only whether the object satisfies the formula $\phi(x)$. Then we *still* get a contradiction. Or at least we do so if our formalism provides us with a notation to express "satisfying." For we can then consider the formula $\phi(x)$ symbolizing "x is a formula with one free variable that does not satisfy itself" and ask whether this formula satisfies itself. A Russell-style contradiction will be forthcoming.

The ordinary-language version of this contradiction is known as the *Grelling* or *heterological* paradox. It is most simply formulated for adjectives (or adjectival phrases). Call an adjective *autological* if, like "short" and "polysyllabic" and "English," it is true of itself, and *heterological* if, like "long" and "monosyllabic" and "French," it is not. We may then ask whether the adjective "heterological" is heterological. (Or if the fact that "autological" and "heterological" are neologisms is distracting, we may ask whether the adjectival phrase "not true of itself" is true of itself.) Paradox arises even if we restrict our attention just to words and phrases that have actually been written down—even if we restrict our attention to concrete inscriptions made of ink or chalk. Contemplation of the parallel between Russell's set-theoretic paradox and Grelling's semantic paradox suggest that it is not "abstract ontology" but a kind of "self-referentiality" or "vicious circularity" that is to blame. In the course of trying to reconstruct the program of Frege's *Grundgesetze*, and his own (1903), Russell was led to introduce a distinction between *predicative* and *impredicative* specifications, and to ban the latter as viciously circular.

Let me now try to characterize as simply as possible Russell's extraordinarily complex system. The first thing that needs to be said is that Russell characterized his theory as a "no classes" theory, and it was so in a double sense. First, Russell dropped what *Frege* called "classes," namely, extensions of concepts, which I have been calling "sets." But second, Russell also dropped what *he* called "classes," which nearly enough were the concepts themselves. For if Russell did not mean by "class" exactly what Frege meant by "concept," he at any rate ascribed to what he understood by "classes" exactly the same features that Frege ascribed to what he understood by "concepts": they are supposed to come in levels, and to be governed by the laws of comprehension and extensionality. For concepts Russell substituted entities of a different sort as the range of his higher-order variables, entities he called "propositional functions." These were also supposed to come in levels and be governed by an axiom of comprehension. To be sure, Russell thought it necessary to subdivide the levels, and to restrict the axiom—it is in this connection that the issue of predicativity *vs* impredicativity comes to the fore—but a Fregean might have thought so, too, after the paradoxes. The crucial difference between "propositional functions" and "concepts" lies not in the subdivision and restriction, but in the fact that the former are not, as the latter are, governed by an axiom of extensionality.

Thus there are in all *three* restrictions imposed on Frege's system. First, the extension operator ‡, converting a higher-level entity X into a zeroth-level entity $x = ‡X$, is dropped. Second, the comprehension axiom is restricted. Third, the axiom of extensionality is dropped. So far as avoiding the paradoxes is concerned, this triple retrenchment is overkill, and with three hands tied behind his back Russell unsurprisingly found himself unable to develop arithmetic within his system without certain additional, as it were compensatory, assumptions. For this and other reasons, Russell's system not only is missing three components present in Frege's, but has present three assumptions with no counterpart in Frege's system, called the axioms of *reducibility*, of *infinity*, and of *choice*.

The resulting theory of propositional functions is inordinately complicated, even compared with Frege's theory of concepts. Moreover, Russell was not nearly so clear and explicit as Frege had been about just what assumptions he was making, so that there are scholarly debates to this day about exactly what Russell's theory amounted to. Fortunately, no more than a brief summary of one plausible interpretation of the Russellian ramified theory of types will be necessary for present purposes.[18] In presenting the summary, I will consider first the difficulties Russell got himself into by dropping extensionality, then those arising from his restricting comprehension and dropping extensions, in each of the three cases indicating how Russell proposed to extricate himself.

To begin with extensionality, for Frege the higher-order variables ranged over concepts, which were supposed to be the sort of things that the referent of a predicate would be, namely, items with gaps that when filled in produced truth-values, which Frege took to be the referents of sentences. For Russell the same variables ranged over propositional functions, which were more like the *senses* of predicates, in that they were items with gaps that when filled in produced propositions, which are much like what Frege took to be the *senses* of sentences. Since predicates may have different senses though applying to the same entities, it was entirely reasonable to repudiate the axiom of extensionality when switching from concepts to propositional functions. Working with propositional functions rather than concepts or classes is inconvenient, however, mainly because without extensionality one is hardly ever in a position to speak of "the" item fulfilling a given condition.

Now as it happens, if one is not going to go beyond the second-order level then one does not need to assume extensionality as an axiom, because one already has it as a theorem. More precisely, if one takes $X \equiv Y$ not as primitive but as an abbreviation for $\forall x(Xx \leftrightarrow Yx)$, as I have been doing, then each instance of extensionality in the following form will be deducible, provided ϕ contains no second-level or higher concept variables:

(1) $$X \equiv Y \rightarrow (\phi(X) \leftrightarrow \phi(Y))$$

This is because then X can appear in $\phi(X)$ only in atomic formulas of the form Xv, and Y only in corresponding atomic formulas of the form Yv, and we have $Xv \leftrightarrow Yv$ for all v. Extensionality is still deducible if one adds, say, a function symbol † applying to arguments of sort X and giving values of sort x, provided one has $X \equiv Y \rightarrow \dagger X = \dagger Y$. And extensionality is *still* deducible if one adds a second-level constant **A** so long as one has $X \equiv Y \rightarrow \mathbf{A}X \leftrightarrow \mathbf{A}Y$. It is for this reason that an *axiom* of extensionality was not actually required by Frege.

With third-order logic and variables for second-level concepts, however, extensionality is *not* a theorem. That is, one can*not* deduce the following:

$$\forall \mathbf{Z}(X = Y \rightarrow (\mathbf{Z}X \leftrightarrow \mathbf{Z}Y)) \tag{2}$$

And Russell was obliged to make much more use of third- and even fourth-order logic than Frege ever did, so for him the absence of an extensionality axiom was a genuine inconvenience. But Russell found a way around this inconvenience.

He showed that his theory with the addition of extensionality becomes true if the formulas of the language are suitably reinterpreted. The reinterpretation involves what is known as *relativization* of quantifiers. In general, relativizing quantifiers to a condition $\theta(u)$ means replacing quantifications $\forall u(\ldots)$ and $\exists u(\ldots)$ by $\forall u(\theta(u) \rightarrow \ldots)$ and $\exists u(\theta(u) \ \& \ \ldots)$. In the present instance, there is no need to change quantifiers involving zeroth-level or first-level variables $x, y, z, \ldots$ or $X, Y, Z, \ldots$ But quantifications over second-level concepts are to be relativized to the condition of being *extensional*, meaning not making distinctions among coextensive first-level concepts. Thus $\forall \mathbf{Z}(\ldots)$ is understood as meaning the following:

$$\forall \mathbf{Z}(\forall X \forall Y(X = Y \rightarrow (\mathbf{Z}X \leftrightarrow \mathbf{Z}Y)) \rightarrow \cdots) \tag{3}$$

Obviously, with the quantifier in (2) thus reinterpreted, (2) becomes deducible.

The treatment of third- and higher-level cases and of many-place cases is more complicated, but the details are unimportant here. The final result is that so long as one is only interested in extensional propositional functions—and Russell held that those were the only ones of relevance to mathematics—one may as well assume the axiom of extensionality. What one then says won't be literally true, by Russell's lights, but will become true if quantifiers are tacitly understood as appropriately relativized.

This being understood, for the remainder of this exposition I will revert to the Fregean language of concepts when speaking of the Russellian theory of propositional functions. And after we are done with Russell, we are hardly ever going to go above the level of second-order logic, so almost always *extensionality will be available "for free,"* for the reason explained above.

TURNING NEXT TO RESTRICTIONS on comprehension, the instance that leads to Russell's paradox reads as follows:

$$\exists Y \forall x (Yx \leftrightarrow \exists X(x = \ddagger X \;\&\; {\sim}Xx)) \tag{4}$$

Here the concept Y being asserted to exist corresponds to a formula involving quantification over all concepts, *Y itself included.* Russell considered this kind of specification of an entity by a condition involving quantification over all entities of the same kind as itself to be viciously circular. The particular kind of circularity such a specification exhibits, whether one considers it vicious or not, is called *impredicativity*. Russell proposed to ban it. Actually, for Russell, (4) is already banned because he does not have the operator ‡, but now it is to be banned again by his predicativity restriction on the axiom of comprehension.

To be more explicit about what Russell *did* assume, he imposed a restriction on ϕ in the axiom of comprehension:

$$\exists X \forall x (Xx \leftrightarrow \phi(x)) \tag{5}$$

Russell assumes (5) only for ϕ that are *predicative* in the sense of containing no first-level concept variables $X, Y, Z, \ldots$ or relational concept variables $R, S, T, \ldots$ that are bound by quantifiers.[19] An analogous assumption is made for relational concepts (of however many places). Note that both free and bound zeroth-level variables $x, y, z, \ldots$ as well as free first-level concept and/or relational concept variables $X, Y, Z, \ldots$ or $R, S, T, \ldots$ are allowed to appear in the formula ϕ in (5).

One crucial observation is that the instances of comprehension for relational concepts needed to prove that $\approx$ is an equivalence are predicative. What has to be done is the following: first, prove for reflexivity that there exists an R such that $X \approx_R X$; second, prove for symmetry that if there exists an R such that $X \approx_R Y$, then there exists an S such that $Y \approx_S X$; and third, prove for transitivity that if there exists an R such that $X \approx_R Y$, and there exists an S such that $Y \approx_S Z$, then there exists a T such that $X \approx_T Z$. One uses the *identity* on X to get the correspondence R needed for reflexivity, the *inverse* to the given correspondence to get the correspondence S needed for symmetry, and the *composition* of the two given correspondences to get the T needed for transitivity. All three items have predicative definitions, as follows:

Identity	I_X	$= «x, y: Xx \ \& \ y = x»$
Inverse (a.k.a. *Converse)*	R^*	$= «x, y: Ryx»$
Composition	$R \circ S$	$= «x, z: \exists y(Rxy \ \& \ Syz)»$

The kind of concepts admitted so far, which can be specified without any quantification over concepts, may be called *first-round* predicative concepts, because Russell goes on to admit *second-round* predicative concepts, whose specification may require quantification over concepts, but only those of the first round. He then goes on to consider still further rounds. Let us use $X^0, Y^0, Z^0, \ldots$ as variables for first-round predicative concepts, $X^1, Y^1, Z^1, \ldots$ as variables for second-round predicative concepts, and so on.[20] Let us write ϕ^0 for a formula involving no more than free and/or bound object variables and/or free first-round concept variables, and ϕ^1 for a formula involving only these and/or bound first-round concept variables and/or free second-round concept variables, and so on. Then what is assumed in the way of comprehension axioms are the following:

(5a) $$\exists X^0 \forall x(X^0 x \leftrightarrow \phi^0(x))$$

(5b) $$\exists X^1 \forall x(X^1 x \leftrightarrow \phi^1(x))$$

And so on. Where only first-round predicative concepts are admitted, we have a *simple* predicative theory; where first- and second-round predicative concepts are admitted, a *double* predicative theory; and where predicative concepts of all rounds are admitted, a *ramified* predicative theory.

When it comes to developing mathematics within Russell's system, predicativity restrictions create difficulties over mathematical induction. For suppose that we have—never mind how—defined zero and successor. Frege's approach more or less amounts to defining an object to be a natural number if and only if it falls under all *inductive* concepts, where a concept is inductive if zero falls under it and the successor of any object falling under it falls under it. Thus the form of the definition is this:

$$(6) \qquad NN(x) \leftrightarrow \forall X(X0 \ \& \ \forall y(Xy \rightarrow Xy') \rightarrow Xx)$$

Then for any formula $\phi(x)$ we get mathematical induction on $\phi(x)$, which is to say the following:

$$(7) \qquad \phi(0) \ \& \ \forall y(\phi(y) \rightarrow \phi(y')) \rightarrow \forall x(NN(x) \rightarrow \phi(x))$$

from the definition (6) and the relevant instance of comprehension (5). In the predicative system, we need to put on superscripts. If we use a definition of natural number like (6) but beginning $\forall X^{17}$, say, then we will get mathematical induction (7) not for arbitrary formulas but only for formulas ϕ^{17} not involving bound variables X^{17}, Y^{17}, Z^{17}, . . . ,because it is only for such formulas that the relevant instances of comprehension, like (5a) and (5b) but beginning $\exists X^{17}$, are assumed. And unfortunately, any formula involving the notion of natural number will involve a forbidden kind of quantification, and mathematical induction will be inapplicable to it—a very serious deficiency in the theory.

To remedy this deficiency, Russell makes a new assumption, the *axiom of reducibility*—actually, it is a scheme, with one axiom for each superscript—according to which any concept of whatever round is coextensive with one of the first round. Since every formula determines a concept of *some* round or other, every formula in fact determines a concept of the *first* round, and we therefore have the following:

$$\exists X^0 \forall x(X^0 x \leftrightarrow \phi(x)) \tag{8}$$

The subdivision of levels into types or "rounds" is in effect undone.

Ramsey therefore proposed simply dropping the predicativity restrictions and assuming unrestricted impredicative comprehension from the beginning. Russell had thought that his indirect procedure, of imposing predicativity restrictions and then undoing them by assuming reducibility, was somehow needed to block the semantic paradoxes like Grelling's heterological paradox. But Ramsey observed that these paradoxes depend on being able to express semantic notions, and that Russell's formalism provides no means of doing so.

The availability of the Weiner-Kuratowski definition of ordered pair permitted another simplification of the system, namely, dropping all relational concepts. Thus one has in the end a simple hierarchy of one-place concepts—usually called "classes"—of levels one, two, three, and so on, in what is generally understood by "the theory of types" today.

FINALLY, THERE IS the matter of infinity. Suppose for a moment that we add to Russell's original system Frege's assumption of the existence of extensions. It will be instructive to consider how the predicativity restrictions save us from paradox. We do indeed have the following instance of comprehension:

$$\exists X^1 \forall x(X^1 x \leftrightarrow \exists Y^0(x = \ddagger Y^0 \ \& \ {\sim} Y^0 x)) \tag{9}$$

If we let A^1 be such an X^1 as in (9), then what we can conclude from (9) is that A^1 cannot be coextensive with any first-round concept. For letting A^0 be any first-round concept and a its extension, (9) tells us that *a* falls under A^1 if and only if it does not fall under A^0. But this conclusion is in itself no contradiction.

This conclusion does, however, conspicuously contradict the axiom of reducibility. We have therefore to choose between that axiom and the axiom of the existence of extensions. Russell made his choice for the former. What will be explored in the next chapter is what happens if one chooses the latter. The answer that will eventually emerge is that, owing to problems over mathematical induction mentioned earlier in this section, one gets only a fragment of arithmetic. But as a matter of fact, with Russell's choice one gets no arithmetic at all without a second extra assumption beyond reducibility.

To begin with, there can be no question of obtaining numbers as *objects*—or *individuals*, as Russell calls the entities at the bottom of the hierarchy of levels—and for Russell a number will have to be, not as with Frege the extension of a second-level concept, but simply the second-level concept itself. The number two is the second-level concept of being a first-level concept under which there falls an object, another, distinct object, but no further object distinct from both of these. (This is why, as mentioned earlier, Russell has to ascend higher in the hierarchy of levels than Frege ever did, especially when he moves on to the theory of real numbers.)

Without the assumption of extensions to give us "logical objects," it will be impossible to prove the Peano postulates for numbers so conceived. The problem is not over induction, but over the *other* Peano postulates. Just which one of them will be unprovable depends on just how the definition of successor is framed, but they certainly cannot *all* be proved, because together they guarantee the existence of infinitely many numbers, and *that* will be impossible to prove. For without "logical objects" it will be impossible for logic to prove the existence of infinitely many—or for that matter, of *any*—objects, and if there are only *n* objects, there will be at most 2^n first-level concepts, and at most 2^{2^n} second-level concepts, and so certainly not infinitely many numbers, if numbers are second-level concepts.

To get round this problem, Russell simply assumes an *axiom of infinity* to the effect that there *are* infinitely many objects. A question arises as to just how this assumption is to be formulated, which is to say, how infinity is to be defined. If one has the natural numbers, one can define X to be finite if and only if X is equinumerous with the natural numbers less than some natural number n. This may be called the "natural" definition. Without presupposing the natural numbers, there are other definitions that can be given.

Dedekind used the definition that X is finite if and only if X is not equinumerous with any Y contained in but not coextensive with X (any Y such that every object falling under Y falls under X, but not conversely). We call X *Dedekind-finite* when this condition is met. Note that in a predicative context there would be no such thing as "the" condition of Dedekind-finitude. Rather, there is first-round Dedekind-finitude (no first-round predicative one-to-one correspondence . . .), second-round Dedekind-finitude (no second-round predicative one-to-one correspondence . . .), and so on.

Russell himself sometimes (though not, as it happens, in connection with the formulation of the axiom of infinity) used another definition. We say that R *linearly orders* (the objects falling under) X if for all objects x , y, z falling under X we have the following:

Irreflexivity	$\sim Rxx$
Transitivity	$Rxy \,\&\, Ryz \rightarrow Rxz$
Trichotomy	$Rxy \vee x = y \vee Ryx$

Given a non-empty Y contained in X, an R-least (respectively, R-greatest) object falling under Y would be an object x falling under Y such that for any other object y falling under Y one has Rxy (respectively, Ryx). R is a *well-ordering* (respectively, *reverse well-ordering*) of X if it is a linear ordering of X and for every non-empty Y contained in X there exists an R-least (respectively, R-greatest) element. We may call X *Russell-finite* if there is an R that is both a well-ordering and a reverse well-ordering of X. This notion, too, splits up into many notions in a predicative context.

An issue that had come up in mathematics in the interval between Frege and Russell was that of the *axiom of choice*. The details of the formulation of the axiom need not detain us at this stage. There are in fact many equivalent formulations, one being that for every set there is a well-ordering of it. What is relevant here is that, assuming choice, all the various definitions of finitude can be proved equivalent. Without the axiom, however, some formulations are stronger than others, in that implications from the former to the latter can be proved, but not vice versa. Russell in fact showed that a weak infinity assumption for objects sufficed to obtain the Peano postulates two levels up, for natural numbers construed as second-level concepts.[21] The significance of this result was somewhat diminished by the circumstance that Russell did, in the end, feel obliged to introduce the axiom of choice on account of its being needed for certain developments in higher mathematics.

This will do, for present purposes, by way of summary of Russell's approach. My summary has not done justice to Russell's many positive contributions, but rather has emphasized weakness and drawbacks. If the summary accomplishes nothing else, it should make it unsurprising that mathematicians ultimately preferred to adopt as a framework for modern mathematics not the theory of types but an *axiomatic set theory* that emerged from attempts to make explicit the assumptions underlying Cantor's set theory.

As MATHEMATICIANS TOOK that course, Frege's system was increasingly forgotten, until the recent revival. Enough notions have been introduced in the summary of Frege's and Russell's approaches to make it possible at this point to sketch the directions in which different workers have proceeded. If one looks closely at the contradiction in Frege's system, it will be seen to have *two* features, *each* of which contributes to the paradox. First, Frege is very free in assuming the existence of concepts, and in particular his system allows the introduction of the *Russell concept* «x: $x \notin x$». Second, Frege is very free in assuming the existence of extensions for concepts, and in particular his system allows the introduction of the *Russell set* $\{x: x \notin x\}$. Modified Fregean systems fall into two categories, according to whether they try to block the paradox at the first or at the second step.

Predicativity restrictions block the first step on the road to paradox, ruling out the Russell concept. If one stops at the simple predicative level, it looks as if one has allowed the *Russell formula* $x \notin x$ into the language but not allowed it to determine a concept. For that formula is just an abbreviation of $\sim\exists X(x = \ddagger X \ \& \ Xx)$, which is allowed in the simple predicative language, though comprehension is not assumed for it. Now there is no obvious reason why a predicate or open formula may in some cases, like a proper name or singular definite description, fail to have a *referent*; but it is very hard to see how one could reasonably allow formulas into one's language that one does not allow to have *senses*. If one thinks, like Frege, of the variables X as ranging over referents, it seems there is no obvious reason why there *must* be an X for every $\phi(x)$; but if one thinks like Russell of the variables X ranging over senses, there does seem to be a reason to demand an X for every $\phi(x)$. This the ramified predicative theory supplies. In that theory there is no longer a single formula $\sim\exists X(x = \ddagger X \ \& \ Xx)$, but a sequence of formulas, a version with superscript 0 that determines an X with superscript 1, a version with superscript 1 that determines an X with superscript 2, and so on. There is no formula that does not determine a concept, but there is no Russell formula. In either case, whether one stops at the simple level or goes on to ramify, there is no Russell concept «x: $x \notin x$».

An alternative would be to allow full, impredicative comprehension, but block the step from the Russell concept to the Russell set, by restricting the assumption that there exists an extension for every concept. One obvious way to implement this idea formally would be to go back to the kind of presentation of Frege's set theory that has an explicit predicate $€xX$ for "is the extension of". We may retain the axiom of set extensionality characterizing what extensions are like when they exist, namely:

$$€xX \;\&\; €yY \rightarrow (x = y \leftrightarrow X \equiv Y) \tag{10}$$

By itself (10) only implies that extensions are unique when they exist. Frege got into trouble by assuming the axiom of set comprehension $\exists x\, €xX$ asserting that extensions always do exist. The obvious way to restrict this unacceptable axiom would be to replace this unconditional existence axiom by a conditional or biconditional version of one of the following two forms:

$$\sigma(X) \rightarrow \exists x\, €xX \tag{11a}$$

$$\exists x\, €xX \leftrightarrow \sigma(X) \tag{11b}$$

Here $\sigma(X)$ would express a proviso, not met by the Russell concept, taken to be sufficient (in the conditional case) or necessary and sufficient (in the biconditional case) to make it safe to assume the concept X has an extension.

There is also a less obvious way to think about restricting the assumption of the existence of extensions. For we have seen that the assumption of the existence of extensions is equivalent to the assumption of the existence of abstracts for all equivalences, the case of coextensiveness $\equiv$ subsuming the case of any other equivalence E. An alternative to assuming safe extensions would be to assume no extensions at all, but only abstracts for some other equivalence or equivalences. Actually, this *abstractionist* approach more or less subsumes the safe-extensions approach, at least in the biconditional version

(11b). For corresponding to any safety condition $\sigma(X)$ we have an equivalence $\equiv_\sigma$ of *safe coextensiveness*, defined as follows:

$$(12) \qquad X \equiv_\sigma Y \leftrightarrow X \equiv Y \vee (\sim\sigma(X) \;\&\; \sim \sigma(Y))$$

This $\equiv_\sigma$ agrees with coextensiveness for safe concepts, but lumps all unsafe ones together. Assuming abstracts for $\equiv_\sigma$ is equivalent to assuming (11b) plus the existence of one additional nameable object o, distinct from all extensions: the abstract of X with respect to $\equiv_\sigma$ amounts to the extension of X if X is safe, and to o otherwise.

Recent writers have considered both predicative approaches, simple and ramified, and impredicative abstractionist approaches, including some that do not and some that do more or less amount to safe-extensions approaches. In chapter 2 below, I will discuss predicativist modifications of Frege's system and the work of Richard Heck, A. P. Hazen, and others. In chapter 3, I will turn first to the particular abstraction theories of Crispin Wright and his school, then to the general abstraction theory of Kit Fine, and to the safe-extensions approach as it has been pursued along certain lines by George Boolos and as it might be pursued along other lines. In the work of all the authors named, one encounters limits to how much of classical mathematics one can develop (or at any rate, to how much one can develop without resort to *ad hoc* hypotheses). My main goal will be to make clear what the limits encountered are on each approach.

1.5 Mathematical Targets

But first I must say something about the scale by which one measures the scope and limits of a given approach. The discussion of this scale will take the form of a survey of stronger and stronger theories of arithmetic, analysis, and set theory that have been developed by mathematical logicians in the period from Russell to the present. Before launching into the survey, a word needs to be said about what historically was the situation in foundations of mathematics in the period between the publication of Whitehead and Russell (1910–13) and the appearance of Gödel's incompleteness theorems.

Briefly put, the situation was this. There were several proposals for codifying classical mathematics, either in a system of axiomatic set theory or in a system like Russell's. But there were also several philosophical points of view from which these systems, and with them much of classical mathematics, were unacceptable. Some of these restrictive philosophies were stricter than others, the strictest of all being called *finitism*, because it totally repudiated the "actual infinite." David Hilbert, the leading mathematician of the period, put forward a program of trying to provide indirect justification for classical mathematics from a finitist standpoint. It was a clever program, and though doomed to failure, it will be worth reviewing.

One way in which a theory not directly justifiable from a given point of view may be indirectly justified is by *interpretation* of the given theory in another that *is* directly justifiable. Here what one means by an *interpretation* of a theory T_1 in a language L_1 in a theory T_2 in a language L_2 is, roughly speaking, a mapping of the formulas of L_1 into formulas of L_2 with two properties: first, that every axiom of T_1 is mapped to an axiom or theorem of T_2; second, that the mapping preserves logical structure. Together these properties imply what we really want, namely, that every *theorem* deducible from the axioms of T_1 is mapped to a theorem deducible from the axioms of T_2. For axioms go to axioms, and preservation of logical structure implies preservation of logical deducibility. (The "rough speaking" in the foregoing characterization lies in my not having given a precise characterization of just what "preserving structure" requires.)

Philosophically, interpretation is a way of legitimizing a theory that is not legitimate when taken literally. If T_1 is legitimate from a certain philosophical point of view, and T_2 is interpretable in T_1, then deducing a result in T_2 and translating it from the language L_2 to the language L_1 may be regarded as a legitimate way of deducing a consequence of T_1, perhaps in a more efficient or more perspicuous way than it could be deduced working with T_1 directly. Thus through interpretation T_2 may be legitimized indirectly, when taken as a *façon de parler*, even if it cannot be legitimized directly, when taken literally.

Any abbreviated notation provides a trivial example of interpretation. I have said that various notations, beginning with $\neq$, are "unofficial abbreviations" introduced by certain definitions, in this case the definition of $x \neq y$ as $\sim(x = y)$. In this case the language and theory I am in practice using has in it the notation $\neq$ and the assumption of the equivalence of $x \neq y$ with $\sim(x = y)$; the language I in principle regard as the official one does not. But the unofficial language is interpretable in the official language by going through and replacing every occurrence of $x \neq y$ by $\sim(x = y)$. Here unpacking the definition counts as an interpretation of the one theory in the other. A less trivial example is provided by Russell's treatment of extensionality. He gives an interpretation of what he considers the theory of "classes," or type theory *with* extensionality, in his theory of "propositional functions," or type theory *without* extensionality. In this case the interpretation consists in appropriately relativizing quantifiers, as we have seen. Typically interpretations involve both devices mentioned so far: finding definitions by which the notions of the theory being interpreted could be introduced as abbreviations into the theory in which it is being interpreted, and also relativizing quantifiers.

There is also another kind of indirect justification besides interpretation that was of special importance to Hilbert. Suppose we have a language L_1, a theory T_1 in that language, a language L_2 extending L_1, and a theory T_2 in that language extending T_1. If Γ is some class of formulas of L_1, then T_2 is called *conservative* with respect to Γ over T_1 when every formula of class Γ that is a theorem of T_2 is already a theorem of T_1.

Philosophically, conservativeness is a relationship that may indirectly legitimize a theory. If T_1 is legitimate from a certain philosophical point of view, and T_2 is conservative with respect to Γ over T_1, then deducing a result of class Γ in T_2 may be regarded as a legitimate way of deducing a consequence of T_1, perhaps in a more efficient or more perspicuous way than it could be deduced working with T_1 directly. Thus through conservativeness T_2 may be legitimized indirectly, when taken as an instrument of discovery, even if it cannot be legitimized directly. (How *important* the instrument is will depend on how broad the class Γ is, and how much shorter or easier indirect deductions are than direct ones.)

To give at least one example to illustrate the notion, suppose we have a theory T in which we can prove $\exists x\phi(x)$. Then the theory T' obtained by adding a constant c with the axiom $\phi(c)$ is a conservative extension. For any conclusion ψ not mentioning c that is deducible in T' using the axiom $\phi(c)$ will be—by one of the usual laws of first-order logic—deducible in T using only the theorem $\exists x\phi(x)$. This is a rather trivial case, but even in this case T' may not be directly interpretable in T. (It is, by what has already been said, if T proves not only $\exists x\phi(x)$ but $\exists! x\phi(x)$; but this need not be so.)

Let us abbreviate "interpretable in a conservative extension of (with respect to some significant class Γ of formulas)" to "reducible to." Then in general a theory reducible to a theory that is directly acceptable from a given philosophical standpoint should be indirectly acceptable, for a certain kind of instrumental use that has been roughly described, from that standpoint. Or more precisely, this will be so provided the fact of reducibility can itself be established *by acceptable means*, which is to say, by means that are acceptable from the standpoint of the philosophical view in question. Hilbert's idea for defusing finitist criticism of classical mathematics was to look for a proof *by finitist means* that classical mathematics is reducible to finitist mathematics. Owing to certain special features of the situation, the details of which need not concern us here, proving all this turns out to require only proving *by finitist means* that some formal system codifying classical mathematics is *consistent*.[22] Hilbert's program was to look for a finitist proof of the consistency of classical mathematics in this sense.

Gödel derailed the program by showing that given a theory T_2 (perhaps put forward as a formalization of classical mathematics), it cannot be proved by means formalizable in T_2, let alone any weaker theory T_1 (such as might be put forward as a formalization of finitist mathematics), that T_2 is consistent (unless T_2 is inconsistent, in which case *anything* can be proved by means formalizable in T_2). This is not to deny that one can in many instances show by finitist or even still more restricted means that an ostensibly stronger theory in an ostensibly richer language is actually reducible to an ostensibly weaker theory in an ostensibly poorer language; for in fact many such reducibility results do hold. But it is to deny that the gap between the theory that is reduced and the theory to which it is reduced can ever be as wide as that between classical and finitist mathematics.

When reducibility is provable, one obtains with it a proof of *relative* consistency, which is to say, a proof that *if* T_1 is consistent, *then* T_2 is consistent. Rather surprisingly, Harvey Friedman has shown that, under mild hypotheses, the converse is true, which is to say that if one has relative consistency, then in quite general circumstances it follows that one has reducibility as well (and if the theories involved are both finitely axiomatizable, outright interpretability).[23]

Relative to a given *metatheory*—the weak theory in which we are proving our interpretability and conservativeness and related results, about which there is no need to be too specific here—we say T_1 is of *no greater consistency strength* than T_2 if the consistency of T_1 relative to T_2 can be proved. In this case, according as the consistency of T_2 relative to T_1 can or cannot also be proved, we say T_1 is of *equal consistency strength* to or of *strictly less consistency strength* than T_2. Similar definitions could be made for reducibility, but by Friedman's results reducibility strength and consistency strength would coincide in almost all cases.

Now it is a very striking fact that though in principle one could have pairs of theories T_1 and T_2 that are of *incomparable consistency strength*, neither being provably consistent relative to the other, and though one can indeed contrive artificial examples of incomparability, in practice virtually all theories that it is natural to consider and that workers in foundations have actually considered turn out to be comparable. There is a fundamental series of theories of greater and greater consistency strength, and virtually every theory of interest in foundations of mathematics turns out to be of equal consistency strength with one or another theory in this series.

What can be done in the way of partial realization of Hilbert's program is to show in many cases that two theories, one ostensibly much stronger, perhaps in a language ostensibly much richer, than the other, are none the less both of equal consistency strength with the same theory in the fundamental series, and so with each other. The fundamental series provides a scale by which the degree of success of a foundational program may be measured. If the program produces a theory T to which theories fairly high up on the fundamental series can be reduced, then it is correspondingly fairly successful, at least by one important measure of success.

A survey of this fundamental series or scale follows. The first part of the survey, concerned with systems of arithmetic, is relevant to the projects considered in chapter 2. The second part, concerned with systems of analysis and set theory, is relevant to the projects considered in chapter 3, and reading of it therefore may be postponed until after chapter 2. In any case, the reader may do well to skim the material on first reading, and refer back to it as needed.

The authoritative compendium of results about the lower portion of the fundamental series is the work of Hájek and Pudlak (1998). For the middle portion, the authoritative source is Simpson (1999). For the upper reaches, there is Kanamori (1997). A non-technical—or at any rate, only semi-technical—account, intended for an audience of philosophers, of the whole of the series from bottom to top has been given in a lecture series by Friedman, but unfortunately has not yet been reduced to published form. My own account is largely dependent on the three books cited (to which the reader is referred for more detailed explanations and bibliographical citations), and especially on Friedman. The usual disclaimer applies: none of the authors mentioned is responsible for any errors that may have crept in. I will divide the tour of the fundamental series into five stages: the bottom of the scale, systems of arithmetic, systems of analysis, type theories *vs* set theories, and large cardinals.

1.5a *The Bottom of the Scale*

Let us begin with theories of arithmetic, first-order theories whose variables x, y, z, . . . are intended to range over natural numbers. The non-logical symbols present will be a constant 0 for zero, a one-place function symbol $'$ for successor, usually a two-place relation symbol $<$ for order—though in the literature this symbol is often omitted from the official language, and taken to be an unofficial abbreviation—two-place function symbols $+$ and $\cdot$ for addition and multiplication, and sometimes additional such symbols $\uparrow$, $\Uparrow$, . . . for exponentiation, superexponentiation, and so on. There will be the usual logical operators, but we also introduce as abbreviations *bounded* quantifiers, writing

$$\forall y < x\,(\ldots) \quad \text{for} \quad \forall y\,(y < x \rightarrow \ldots)$$
$$\exists y < x\,(\ldots) \quad \text{for} \quad \exists y\,(y < x\ \&\ \ldots)$$

And letting $y \leq x$ abbreviate $y < x \vee y = x$, we similarly use $\forall y \leq x$ and $\exists y \leq x$. Terms built up using function symbols may also appear as bounds. A formula ϕ is called *bounded* or Δ_0 or Σ_0 or Π_0 if all quantifiers (if any) in ϕ are bounded.

The weakest system we will consider is generally called *Robinson arithmetic* or Q. Its axioms are as follows:

(Q1) $\sim x' = 0$
(Q2) $x' = y' \rightarrow x = y$
(Q3) $y = 0 \vee \exists x\,(y = x')$
(Q4) $x + 0 = x$
(Q5) $x + y' = (x + y)'$
(Q6) $x \cdot 0 = 0$
(Q7) $x \cdot y' = (x \cdot y) + x$

For certain purposes, it is useful consider a variant theory, ostensibly stronger but really equivalent, that adds one more primitive, the order symbol <, to the language, adds two more axioms to the theory, and replaces (Q3) by an ostensibly stronger version:

(Q1*) $\sim x < 0$
(Q2*) $x < y' \leftrightarrow x < y \vee x = y$
(Q3*) $y = 0 \vee \exists x(x < y \;\&\; y = x')$

Let us call this variant Q*.[24]

None of the usual associative, commutative, or distributive laws for addition and multiplication can be proved in Q (or Q*) nor can even the law $x' \neq x$. As to this last point, a natural model of Q is provided (as Saul Kripke pointed out to the author) by the cardinal numbers, with x' defined as $x + 1$. For infinite cardinals we then have $x' = x$. With even the law $x' \neq x$ missing, virtually no mathematics can be done in the system Q. The system was indeed introduced by Robinson *not* in order to do mathematics in it, but rather in an effort to determine the scope of Gödel's first incompleteness theorem.

What is proved in the classic monograph of Tarski, Mostowski, and Robinson (1953) is that the theorem in question applies to any theory in which Q can be interpreted, a fact now often found incorporated in textbook presentations of the Gödel theorems. A large part of the task in the classic monograph was to give a definition of *interpretation* that is at once totally rigorous and maximally general. Maximal generality need not concern us here, but we will be needing a less rough idea than has been given so far of just what has to be done to show Q is interpretable in a given theory T.

For present purposes, we may think of giving an interpretation of one theory T_1 in another theory T_2 in the following way. First of all, we want to interpret the objects of T_1 as being, or being represented by, certain of the objects of T_2, namely, those for which a specific formula $\delta(x)$ holds. Statements of T_1 about "all x" will be interpreted in T_2 as statements about "all x such that $\delta(x)$."

Second, each relation symbol P in the language of T_1 must be given a definition in T_2. For instance, if P has two places, we take some specific formula with two free variables $\pi(x, y)$ and introduce Pxy as an abbreviation for it, and similarly for relation symbols of any number of places. And the definitions are supposed to be such as to make the interpretation of any axiom of T_1 come out a theorem of T_2. For instance, if it is an axiom of T_1 that $\forall x \exists y \mathrm{P}xy$, then the following must be a theorem of T_2:

$$\forall x(\delta(x) \rightarrow \exists y(\delta(y) \;\&\; \pi(x, y)))$$

Third, if there are function symbols, these also have to be defined, and in such a way that the interpretation of any axiom of T_1 comes out a theorem of T_2. We have seen how a zero-place function symbol or constant c can be introduced as an abbreviation. We take some specific formula $\gamma(x)$ for which $\exists! x\gamma(x)$ can be proved, and give that x the name c. Similarly, for a one-place function symbol f, we take some specific formula $\phi(x, y)$ for which $\forall x \exists! y \phi(x, y)$ can be proved, and give that y the name $f(x)$. But some caution is needed here. Back in T_1, c was supposed to be among the objects of that theory; so over in T_2, c should be among those objects that are representing objects of T_1, or in other words, those objects for which $\delta(x)$ holds. Thus we need it to be the case that $\delta(c)$ holds. Similarly, we need it to be the case that $\forall x(\delta\ (x) \rightarrow \delta(f(x)))$ holds. As is said, the objects for which $\delta(x)$ holds must be *closed under f*. A similar closure condition must hold for all function symbols, of all numbers of places.

Turning now to the particular theory Q*, to give an interpretation we must specify a formula δ, define all the relevant arithmetical notions (0, $'$, $<$, $+$, $\cdot$), and insure that two lists of conditions hold. First, we must have all the relevant closure conditions, and second, we must have that all the axioms of Q* hold for the objects for which δ holds:

(C0)	$\delta(0)$
(C′)	$\delta(x) \rightarrow \delta(x')$
(C<)	$\delta(x)\ \&\ y < x \rightarrow \delta(y)$
(C+)	$\delta(x)\ \&\ \delta(y) \rightarrow \delta(x + y)$
(C·)	$\delta(x)\ \&\ \delta(y) \rightarrow \delta(x \cdot y)$
(R1)	$\delta(x) \rightarrow 0 \neq x'$
(R2)	$\delta(x)\ \&\ \delta(y) \rightarrow (x' = y' \rightarrow x = y)$
(R1*)	$\delta(x) \rightarrow \sim x < 0$
(R2*)	$\delta(x)\ \&\ \delta(y) \rightarrow (x < y' \leftrightarrow x < y \vee x = y)$
(R3*)	$\delta(x) \rightarrow (x = 0 \vee \exists y(y < x\ \&\ x = y'))$
(R4)	$\delta(x) \rightarrow x + 0 = x$
(R5)	$\delta(x)\ \&\ \delta(y) \rightarrow x + y' = (x + y)'$
(R6)	$\delta(x) \rightarrow x \cdot 0 = x$
(R7)	$\delta(x)\ \&\ \delta(y) \rightarrow x \cdot y' = (x \cdot y) + x$

Let us call a formula $\delta(x)$ *inductive* if (C0) and (C′) hold; *orderly* if also (C<) holds; *additive* if also (C+) holds; and *multiplicative* if also (C·) holds. With this jargon, what we have to do to interpret Q* and hence Q in a theory is introduce the arithmetical notions (0, ′, <, +, ·) as abbreviations, and then find a multiplicative formula $\delta(x)$ such that each axiom of Q* holds for all objects for which δ holds.[25] This is the rather complicated sufficient condition for interpretability of Q that will be used for the proof of various results in the next chapter.

A SIMILAR DEFINITION could be made for interpretation of any other arithmetic theory. It turns out that many ostensibly stronger arithmetic theories than Q are interpretable in Q, without any need to change the definitions of the arithmetical notions (0, ′ <, +, ·), but just by relativization of quantifers to some suitable multiplicative formula $\delta(x)$. Robert Solovay, in unpublished work, showed how the extension of Q obtained by adding the usual associative, commutative, and distributive laws can be interpreted in Q. Nelson (1986) applied the same method to many further laws.

To describe more precisely the scope of Nelson's work, some definitions will be required. To begin with, the reason various laws cannot be directly proved in Q is that their proofs require mathematical induction. The theory PA or P^1 of *first-order Peano arithmetic* adds to Q the axiom scheme of mathematical induction:

(I) $$(\phi(0) \,\&\, \forall x(\phi(x) \rightarrow \phi(x'))) \rightarrow \forall x \phi(x)$$

Any law that could be proved by mathematical induction from the axioms of Q* can be proved in P^1. Nelson showed that for any finite number of laws of a certain kind that could be proved by mathematical induction from the axioms of Q*, we can get an interpretation in Q of Q* plus those laws. The kind of laws in question are those for which the formula $\phi(x)$ on which we are doing induction is not too complicated.

Just what is meant by "not too complicated" is best illustrated by example. Consider how one might try to deduce the transitivity of the order relation <, which is to say the following law:

$$\text{(T)} \qquad z < y \;\&\; y < x \rightarrow z < x$$

The proof would go by induction on x. At the base or zero step, (T) for $x = 0$ is vacuously true by (Q1*). At the induction or successor step, assuming as induction hypothesis (T) for x, we can prove (T) for x' as follows. If $z < y$ and $y < x'$, then by (Q2*) there are only two cases. Either $y < x$, in which case $z < x$ by the induction hypothesis, or else $y = x$ and $z < x$ directly. In either case, since $z < x$, we have $z < x'$ from (Q2*), to complete the proof. Here the formula on which one is doing induction—the ϕ for which one would need the induction axiom (I) above—is the quantifier-free formula $\tau(x, y, z)$ displayed in (T). It is important to note that, even though in the end one is proving a universal generalization $\forall x \forall y \forall z \tau(x, y, z)$, one is not doing induction on $\forall y \forall z \tau(x, y, z)$. That is to say, in showing for particular values of the parameters y and z that (T) holds for x' if it holds for x, one does *not* use the assumption that (T) holds for x for any *other* values of the parameters y and z. The proof of (T) is a proof by *quantifier-free* induction. What Nelson showed is that for any laws $\alpha_1, \ldots, \alpha_n$ that could be proved by quantifier-free induction, the proof for α_i perhaps using α_j for $j < i$, Q* plus the α_i can be interpreted in Q. In fact, he pushed this further to allow a more general kind of induction, namely for bounded or Δ_0 formulas.

The theory consisting of Q or, equivalently, Q* plus all Δ_0-induction axioms is called in the literature $I\Delta_0$ or $I\Sigma_0$ or $I\Pi_0$ by some and PFA for *polynomial functional arithmetic* by others. Here I will call it Q_2 or *Nelson-Wilkie arithmetic*.[26] Nelson's result, just described, is sometimes called the *local* interpretability of this theory in Q. Alex Wilkie proved the *global* interpretability of this theory in Q. That is, he showed that one can get an interpretation in Q in the general sense of interpretation I have discussed above, namely, one that will give *all* instances of (I) for Δ_0 formulas—all Δ_0-induction axioms—at once, and hence *all* laws provable by Δ_0-induction in a *single* interpretation.

The results I have described about interpretability in Q show that Q and Q_2 are at the same level in the scale of theories alluded to earlier, a level that for our purposes will be the bottom level. Already at this level much more can be done than has been described—Nelson and Wilkie already both got other laws besides ones that are theorems of Q_2, and more has been done since—and a significant amount not just of arithmetic but of algebra and analysis can be done. So establishing that Q is interpretable in a given theory *T* is showing that a significant amount of mathematics is interpretable in *T*. Nonetheless, it is still a long way up from this bottom level even to the level of first-order Peano arithmetic.

1.5b *Systems of Arithmetic*

More can be done if we add more operations from the series $'$, $+$, $\cdot$, $\uparrow$, $\Uparrow$, ... in which each operation is obtained by recursion from the one before (addition is repeated taking of successors, multiplication is repeated addition, and so on). We get the system I will call Q_3 by adding $\uparrow$ with the recursion equations for exponentiation, namely:

(Q8) $\qquad x \uparrow 0 = 0'$

(Q9) $\qquad x \uparrow y' = (x \uparrow y) \cdot x$

Q_3 is a version of what has been called *Kalmar arithmetic*, nowadays usually called $I\Delta_0(\exp)$ or sometimes EFA for *exponential functional arithmetic*. Note that on adding the symbol $\uparrow$ to the language, we get new formulas, including new bounded formulas, and hence new instances of Δ_0-induction. Using these instances, the usual laws of exponents can be proved.

Actually, introducing the symbol $\uparrow$ is in a sense optional. In the lemmas leading up to Gödel's theorems in modern textbook presentations, it is shown that even without adding any new symbols there is already a formula $\theta_\uparrow(x, y, z)$ that "represents" exponentiation in Q, in a sense implying that if $a \uparrow b = c$, then Q can prove the following:

$$\forall y(\theta_\uparrow(\mathbf{a}, \mathbf{b}, y) \leftrightarrow y = \mathbf{c})$$

Here **a** and **b** and **c** are the formal numerals for *a* and *b* and *c* (namely, **0** is 0, **1** is $0'$, **2** is $0''$, and so on). What Q can*not* prove is the following existence and uniqueness lemma for exponentiation:

(exp) $$\forall x \forall y \exists! z\, \theta_\uparrow(x, y, z)$$

(Existence causes more trouble than uniqueness.) It is not strictly necessary to add the symbol $\uparrow$ and its defining equations (Q8) and (Q9) to treat exponentiation, since it turns out that just adding instead (exp) as a new axiom produces a system in which Q_3 is interpretable. Inversely, there is the option of introducing *more* symbols, such as ! for the factorial function, along with the appropriate recursion equations:

(!0) $0! = 0'$ (!′) $y'! = y! \cdot y'$

For it turns out that the analogue of (exp) for the factorial can be proved in Q_3, permitting ! to be defined as an abbreviation, and the recursion equations to be deduced as theorems. The system with the new symbol and axioms is interpretable in the system Q_3 without it.

Adding superexponentiation, with the symbol $\Uparrow$ and the recursion equations for it, produces Q_4. This might be called

Gentzen arithmetic, not because the proof theorist Gerhard Gentzen made a special study of it, but for another reason I should pause to explain. As is generally indicated in textbook treatments, a proof of Gödel's second incompleteness theorem is obtained by formalizing the proof of the first incompleteness theorem in a theory of arithmetic. Nelson's special interest was in formalizing results about provability in systems weak enough to be interpretable in Q. That is rather tricky, whereas formalizing the proof of the first incompleteness theorem in Q_3 is merely tedious. (Laszlo Kalmar introduced a system like Q_3 precisely in order to a have weak context in which the Gödel proof could be formalized.) Many other proof-theoretic results besides the Gödel incompleteness theorem can be formalized in Q_3 as well. However, Gentzen's key result, the *cut-elimination theorem*, which is the cornerstone of higher proof theory, turns out to require Q_4. Other names for Q_4 are $I\Delta_0$(superexp) or SEFA for *superexponential functional arithmetic*. Again one has the option of omitting the symbol $\Uparrow$ and inversely the option of adding certain other symbols for certain other recursive functions.

Continuing on in this way we get Q_5, Q_6, Q_7, and so on, and their union Q_ω, which for reasons that will become apparent in a moment I will call *Grzegorczyk arithmetic*. Again one has the option of omitting the symbols $\uparrow$, $\Uparrow$, …, or inversely of adding symbols for certain other recursive functions, namely, those for which the analogue of (exp) can be proved. These are, by a result essentially due to Andrzej Grzegorczyk, precisely the *primitive* recursive functions.[27]

A DIFFERENT WAY OF strengthening Q_2 would be not to allow additional operations, but to allow induction for a larger class of formulas. It is a law of general logic that every formula is equivalent to one consisting of a string of quantifiers at the beginning, followed by a quantifier-free formula. In formal systems of arithmetic, *a fortiori* every formula is equivalent to one with a string of quantifiers at the beginning, followed by a bounded or Δ_0 formula. If in such a formula all the quantifiers are existential (respectively, universal), the formula is called Σ_1 (respectively, Π_1). A string of existential (respectively, universal) quantifiers in front of a Π_1 (respectively, Σ_1) formula produces a Σ_2 (respectively, Π_2) formula. And so on.

Instead of allowing induction only for Δ_0 formulas, one may consider allowing induction for Σ_1 formulas, thus producing *Parsons arithmetic*, as I call it, or $I\Sigma_1$. Using Σ_1-induction, it is possible to prove the existence and uniqueness assumption for exponentiation and indeed the analogous assumption for superexponentiation, superduperexponentiation, and so on, and more generally for all primitive recursive functions. Thus Q_ω is interpretable in $I\Sigma_1$. But inversely, $I\Sigma_1$ is reducible to Q_ω, as was proved by Charles Parsons.[28] In particular, one does not get any recursive functions beyond the primitive recursive.

Next $I\Sigma_2$ is defined in the obvious way. It may be called *Ackermann arithmetic*, after Wilhelm Ackermann, who produced a memorable example of a recursive function that is not primitive recursive. It has variant presentations, one of which is as follows. Write $[y]$ for the yth operation in the series $+$, $\cdot$, $\uparrow$, $\Uparrow$, $\ldots$, and consider the function $a(x, y, z) = x\ [y]\ z$. This function is recursive, but not primitive recursive. If one tries to formalize the proof of the existence and uniqueness assumption for this Ackermann function, one finds oneself needing to go beyond Σ_1-induction. The proof is naturally formalized in $I\Sigma_2$.[29]

The series continues in the obvious way with $I\Sigma_3$, $I\Sigma_4$, $I\Sigma_5$, and so on, until—every formula being logically equivalent to a Σ_n-formula for some n—one gets the extension of Q_2 with induction (I) allowed for *all* formulas of the language. This resulting theory is a version of *first-order Peano arithmetic* PA or P^1, already mentioned out of turn above, which usually plays a prominent role in textbook presentations of the Gödel theorems. A great deal of mathematics can be done in P^1.

In particular, though it is ostensibly about natural numbers only, the theory of the integers and rational numbers can be developed in P^1. In Cantor's theory of transfinite cardinals, the smallest, $\aleph_0$, defined as the number of natural numbers, is shown to be also the number of integers, of rational numbers, and of finite strings from a finite alphabet of symbols, among other examples. In each case the proof proceeds by explicitly exhibiting a one-to-one correspondence. The natural number that corresponds to a given integer, rational number, or finite string of symbols may be thought of as a *code number* for that item. By replacing the original items with their code numbers, theories of integers, of rational numbers, and of finite strings of symbols can be interpreted in first-order Peano arithmetic P^1. To obtain stronger theories, in which the theory of the *real* numbers can be developed, we will have to move beyond the language of arithmetic in which all theories considered so far have been formulated.

1.5c *Systems of Analysis*

Another crucial transfinite cardinal in Cantor's theory is $\mathfrak{c} = 2^{\aleph_0}$, defined as the number of sets of natural numbers, and shown also to be the number of real numbers (or points on the Euclidean line) and of complex numbers (or points in the Euclidean plane), as well as the number of infinite sequences $\langle A_0, A_1, A_2, \ldots \rangle$ of sets of natural numbers, or of real numbers, or of complex numbers. While there are more sets of real numbers and functions on real numbers than real numbers, mathematical *analysis*, the branch of mathematics that begins with differential and integral calculus, is generally concerned less with arbitrary sets of and functions on real numbers, than with special ones, such as what are called *open* sets and *continuous* functions, and of these there are no more than there are real numbers. Again each proof proceeds by explicitly exhibiting a one-to-one correspondence, permitting sets of natural numbers to code real numbers, complex numbers, infinite sequences, open sets, continuous functions, and more, and to interpret theories of such items in *second-order* Peano arithmetic PA^2 or P^2, which system therefore has among mathematical logicians the alternate name *analysis*.

P^2 is just like P^1 except that it is a (monadic) *second*-order theory with full, impredicative comprehension, and with induction in the form of a single axiom:

(CA) $\exists X \forall x (Xx \leftrightarrow \phi(x))$

(I_0) $X0 \; \& \; \forall x (Xx \rightarrow Xx') \rightarrow \forall x Xx$

Any instance of the axiom scheme of (I) for any formula ϕ follows from (I_0) by the instance of (CA) for ϕ. There are a number of subsystems of P^2 that are of special interest, and our survey of systems of analysis and set theory may begin with these, before turning to stronger theories. But before describing the systems, the various classes of formulas they involve must be defined.

Formulas with only bounded first-order quantifiers, which were called Δ_0 or bounded in the case of first-order arithmetic, are called Δ_0^0 or again *bounded*. Formulas consisting of a string of existential (respectively, universal) first-order quantifiers followed by a Δ_0^0 formula, which were called Σ_1 (respectively Π_1) in the case of first-order arithmetic, are called Σ_1^0 (respectively, Π_1^0), and similarly for subscripts $n > 1$. The hierarchy of formulas with superscript 0 is called the *arithmetical hierarchy*. Formulas with only first-order quantifiers are called Δ_0^1 or *arithmetical*. Formulas consisting of a string of existential (respectively, universal) second-order quantifiers in front of an arithmetical formula are called Σ_1^1 (respectively, Π_1^1), and similarly for subscripts $n > 2$. The hierarchy of formulas with superscript 1 is called the *analytical hierarchy*. These are the first two levels of the *Kleene hierarchy* of formulas of higher-order arithmetic, and the only two levels that will concern us here. Under minimal assumptions, every second-order formula is equivalent to an analytic formula.

Of special interest are five theories singled out by Friedman, and intensively studied by Simpson and his school. I will simply call the systems F_0, F_1, F_2, F_3, F_4. They are known in the literature as RCA_0, WKL_0, ACA_0, ATR_0, Π^1_1-CA_0. Each replaces (CA) by something weaker, but keeps induction in the form (I_0).[30] Each admits several different but equivalent axiomatizations, which is to say, several different but equivalent replacements for (CA). The simplest to present are the following, wherein ϕ^0_0 stands for any *bounded* formula, and ϕ^1_0 for any *arithmetical* formula, and for $n = 2$ two equivalent alternatives are listed:

(F_0) $\quad \forall x(\exists y\phi^0_0(x, y) \leftrightarrow \forall y\psi^0_0(x, y)) \rightarrow \exists X\forall x(Xx \leftrightarrow \exists y\phi^0_0(x, y))$

(F_1) $\quad \sim\exists x(\forall y\phi^0_0(x, y) \;\&\; \forall y\psi^0_0(x, y)) \rightarrow$
$\quad\quad \exists X(\forall x(\forall y\phi^0_0(x, y) \rightarrow Xx) \;\&\; \forall x(\forall y\psi^0_0(x, y) \rightarrow \sim Xx))$

(F_2a) $\quad \exists X\forall x(x \leftrightarrow \exists y\phi^0_0(x, y))$

(F_2b) $\quad \exists X\forall x(x \leftrightarrow \phi^1_0(x))$

(F_3) $\quad \sim\exists x(\exists Y\phi^1_0(x, Y) \;\&\; \exists Y\psi^1_0(x, Y)) \rightarrow$
$\quad\quad \exists X(\forall x(\exists Y\phi^1_0(x, Y) \rightarrow Xx) \;\&\; \forall x(\exists Y\psi^1_0(x, Y) \rightarrow \sim Xx))$

(F_4) $\quad \exists X\forall x(x \leftrightarrow \forall Y\phi^0_0(x, Y))$

To describe these axioms in words, F_0 asserts the existence of any set that can be defined *both* by a Σ^0_1 *and* by a Π^0_1 formula; which for formulas without second-order parameters turn out to be precisely the *recursive* sets. An alternate name is Δ^0_1-CA_0. In F_1, any Π^0_1 formulas that are disjoint, in the sense that there is nothing for which they both hold, can be *separated* by some set X; in this case, the X asserted to exist is *not* described by any particular formula. An alternate name is Π^0_1-SP_0, with SP standing for "separation principle." F_2 asserts the existence of any set that can be defined by a Σ^0_1 formula, which for formulas without second-order parameters turn out to be precisely the *recursively enumerable* sets; but allowing the existence of such sets turns out to be equivalent to allowing the existence of any set that can be defined by an arithmetical formula, which for formulas without second-order parameters are the *arithmetical* sets, thus giving precisely the simple predicative monadic second-order extension of first-order Peano arithmetic. Alternate names are Σ_1^0-CA_0 and Δ_0^1-CA_0. F_3 is analogous to F_1, and an alternate name is Σ_1^1-SP_0, analogous to Π_1^0-SP_0. F_4 or Π_1^1-CA_0 asserts the existence of any set that can be defined by a Π_1^1 formula.

There are several reasons why these particular systems, out of all the subsystems of P^2 that might be identified, are of special interest. F_0 is already sufficient for a significant amount of analysis (rather as Q_2 is already sufficient for a significant amount of number theory), while F_4 is sufficient for virtually all analysis. In between, if a theorem of analysis is *not* provable in F_0, but has a natural proof in one of the higher F_i, it very often turns out that adding the theorem to F_0 enables one to prove the characteristic axiom of F_i. The practice of "proving axioms from theorems" in this way is called *reverse mathematics*. The fact that such reversals are possible shows that the F_i represent rather natural fragments of full analysis.

For $i > 0$, each F_i corresponds to a certain critical school in the philosophy of mathematics, in the following sense. For each of the schools in question, there is a system that is generally accepted as formalizing just what is *directly* justifiable from the point of view of that school. These systems are *not* first-order theories.[31] But in each case, one of the F_i is *reducible* to the system in question. In particular, for Π_2^0 formulas, F_1 is conservative over F_0, which in turn is conservative over $I\Sigma_1$, which in turn is conservative over Q_ω, and Hilbert's program of reduction to finitist mathematics can be carried out for F_1. F_2 is a conservative extension of P^1 by a result I will be referring to as Shoenfield's theorem.[32] This theory is strictly stronger than F_1 as P^1 is strictly stronger than $I\Sigma_1$. It is also known that F_3 is strictly stronger than F_2, and F_4 strictly stronger than F_3.

BEYOND $F_4 = \Pi^1_1\text{-CA}_0$ there extends a series of stronger and stronger theories $\Pi^1_n\text{-CA}_0$, leading up to full classical analysis P^2. It may be mentioned that though officially we have taken P^2 to be a monadic second-order theory with $+$ and $\cdot$ as primitives, since a coding of pairs can be defined using $+$ and $\cdot$, thus giving us the effect of polyadic second-order logic, we could instead have taken P^2 to be a polyadic second-order theory with just 0 and $'$ and with just the Peano postulates, which is to say, induction plus (Q1) and (Q2), since in this theory $+$ and $\cdot$ can be defined by Dedekind's theorem.

When we ascend to the third-order Peano arithmetic P^3, polyadic second-order logic can actually be simulated by monadic third-order logic, so that we can dispense with $+$ and $\cdot$ as primitives. The series of nth-order arithmetics P^n leads up to P^ω, with nth-order variables for all n, and this is essential the simple theory of types with the axiom of (Dedekind) infinity. The name PM is sometimes attached to this theory, though the actual system of Whitehead and Russell (1910–13) was, as explained in the discussion of Russell's solution to his paradox, rather different.

1.5d *Type Theories* vs *Set Theories*

Turning now from analysis to set theory, the most widely accepted system of axiomatic set theory is called ZFC. It is a first-order theory with a single non-logical symbol, the two-place relation symbol $\in$. There is no symbol for sethood, because it is assumed in the theory that *every* object is a set. Accordingly, the axiom of *extensionality* is formulated as in the top line of table D. The subsystem Z of *Zermelo* set theory has, in addition to extensionality, the next six or seven axioms in the table.

Note that in the first four axioms in the table, the $\exists x$ could be strengthened to $\exists! x$ by extensionality. We then give the usual names to the x involved: $\emptyset$ for the null set, $\{u, v\}$ for the unordered pair, $\bigcup u$ for the (grand) union, $\wp(u)$ for the power set. These axioms imply the existence of the singleton $\{u\} = \{u, u\}$, the ordered pair $\langle u, v\rangle = \{\{u\}, \{u, v\}\}$, the (simple) union $u \cup v = \bigcup \{u, v\}$, the adjunction $u \wedge y = u \cup \{y\}$, and so on. In the statement of infinity the abbreviation $y' = y \wedge y$ has been used.

Separation is so called because it allows us to separate out from any set u those elements for which a condition $\phi(y)$ holds to form a set $\{y \in u\colon \phi(y)\}$. Note that by separation, to get pairing as stated it would be enough to have $\exists x(u \in x \;\&\; v \in x)$. For having such a set x, we would have $\{u, v\} = \{y \in x\colon y = u \vee y = v\}$. Similar remarks apply to union and power. Likewise, using separation and extensionality we can separate out from any set x for which the condition of the axiom of infinity holds the set of all its elements y that belong to *every* set for which that condition holds, to obtain the unique *smallest* set for which that condition holds, called ω.

Foundation, also called *regularity*, was not included among Zermelo's original axioms, though it was accepted by him later; and even today it is not always counted among the official axioms, though usually it is. But it need not detain us at present. The subsystem ZF of *Zermelo-Frankel* set theory adds a further axiom, *replacement*, which in fact admits several different formulations, all equivalent in the presence of the other axioms, and some such as to make separation redundant. For the present we may adopt the version shown in the table. Replacement is so called because it allows us, whenever a condition $\psi(v, w)$ associates to every v some unique $w = v^{\psi}$, to replace all the elements v in a set u by the associated v^{ψ} to form a set $\{v^{\psi}\colon v \in u\}$. The full system ZFC adds the axiom of *choice* AC, which need not concern us at present.

Various subsystems will be of interest. We may write a superscript minus sign to indicate dropping the power set axiom, and $+ \wp(\omega)$ to indicate replacing the general power set axiom by a weaker axiom guaranteeing the existence of the power set of ω. We may write $\wp^2(u)$ for $\wp(\wp(u))$ and similarly $\wp^n(u)$ for $n > 2$, and may use the notation $+ \wp^n(\omega)$. Alternatively, we may write $-\infty$ to indicated dropping the infinity axiom.

We may also introduce as abbreviations *bounded* quantifiers, writing

$$\forall y \in x\,(\ldots) \quad \text{for} \quad \forall y\,(y \in x \rightarrow \ldots)$$
$$\exists y \in x\,(\ldots) \quad \text{for} \quad \exists y\,(y \in x \mathbin{\&} \ldots)$$

Analogous to the arithmetical and analytical hierarchies there is the *Levy hierarchy* of formulas of set theory. Those with only bounded quantifiers are called *bounded* or Δ_0 or Σ_0 or Π_0. Prefixing existential (respectively, universal) quantifiers gives Σ_1 (respectively, Π_1) formulas, and the hierarchy continues with Σ_n (and Π_n) for $n > 1$ in the obvious way. A subscript n on Z will indicate restricting separation to Σ_n formulas ϕ, and a subscript n on ZF will indicate similarly restricting replacement as well.

We will see in the next chapter that $Z^- - \infty$ (and indeed extensionality, null set, adjunction, and separation) is sufficient to give a system in which first-order Peano arithmetic P^1 can be interpreted. Inversely, $Z^- - \infty$ (and indeed replacement and for that matter choice as well) can be interpreted in P^1, using a coding of finite sets by natural numbers. Other pairs of theories are similarly related: Z_n^- and Π^1_{n+1}-CA_0, $Z(F)^-$ and P^2, $Z(F)^- + \wp(\omega)$ and P^3, $Z(F)^- + \wp^n(\omega)$ and P^{n+2}. The parenthetical (F) indicates that we do not need replacement to interpret P^n in set theory, but do get replacement when we interpret set theory in P^ω. *Z* is slightly stronger than the simple theory of types P^ω. ZF is considerably stronger than Z, though one has to look long and hard to find any theorem of mathematical analysis of interest to working mathematicians where the difference between the two systems would make a difference.

Many extensions of ZF by the addition of further axioms turn out to be interpretable in ZF itself. Notably, by results of Paul Cohen, assuming ZF is consistent, AC is not a theorem of it. Nor is the *continuum hypothesis* CH, asserting that the number $\mathfrak{c}$ of real numbers is the next larger cardinal $\aleph_1$ after the number $\aleph_0$ of natural numbers, a theorem of ZFC = ZF + AC. Nor is its generalization GCH a theorem of ZFC + CH. Nor is Gödel's *axiom of constructibility* V = L, whose formulation need not detain us, a theorem of ZFC + GCH. Yet by earlier results of Gödel, ZFC + V = L is interpretable in ZF, and moreover implies GCH, hence CH, as well as AC. These results also apply *mutatis mutandis* to the subtheories of Z and ZF considered in the preceding paragraph.

1.5e *Large Cardinals*

The simple predicative second-order version of ZFC is known as NBG or *von Neumann–Bernays–Gödel set theory*. It is a conservative extension, by Shoenfield's theorem. The second-order theory ZFC^2 is generally known as MK or *Morse-Kelly set theory*. In both the predicative and full second-order versions, the scheme of separation is replaced by a single axiom, thus:

Separation $\qquad \exists x \forall y(y \in x \leftrightarrow y \in u \;\&\; Uy)$

And similarly for replacement. MK is stronger than ZFC, but only slightly so.

More significantly stronger theories arise when one adds to ZFC certain *higher infinity* or *large cardinal* axioms. These assert the existence of some very large sets, thereby implying the existence of some very large cardinal numbers. The formulation of these candidate axioms is in all cases rather technical, though it is least so in the case of some of the weaker ones among them.

A set M is called *transitive* if $y \in M$ whenever $x \in M$ and $y \in x$. A *natural candidate model* for ZFC has a transitive set M as its domain, and the actual elementhood relation $\in$ among elements of M as the denotation assigned to the elementhood symbol $\in$, so that $a, b \in M$ satisfy $x \in y$ if and only if $a \in b$. Transitivity guarantees that extensionality will be true in such a candidate model: if elements a and b of M are distinct, there is some c that is an element of one but not the other of them, and this c will by transitivity itself be an element of M, and a and b and c will satisfy $\sim(z \in x \leftrightarrow z \in y)$ in the model. Generally speaking, we get a model of the other axioms only if M fulfills some additional closure conditions (contains the null set, contains the unordered pair of any sets it contains, and so on).

A set M is called *supertransitive* if M is transitive and if $y \in M$ whenever $x \in M$ and $y \subseteq x$. A *natural candidate model* for $ZFC^2 = MK$ has a supertransitive set M as the domain of its first-order variables, and the power set $\wp(M)$ as the domain of its second-order variables, with $a, b \in M$ satisfying $x \in y$ if and only if $a \in b$, and $a \in M$ and $A \subseteq M$ satisfying Xx if and only if $a \in A$. Supertransitivity guarantees that the single axiom version of separation above will be true in such a candidate model.

ZFC (if consistent) cannot prove that any natural candidate model is indeed a model of ZFC, for that would prove that ZFC has a model, and therefore would prove that ZFC consistent, and no consistent theory can prove its own consistency. Similarly, while ZFC^2 can prove that there is a natural model of ZFC, it cannot prove that any natural candidate model of ZFC^2 is indeed a model of ZFC^2. The hypothesis of the *existence of an inaccessible cardinal* is equivalent to, and for present purposes may simply be taken to *be*, the hypothesis that there exists a natural model of ZFC^2. The cardinal number of such a set is called an *inaccessible* cardinal. The set theorists' usual formulation of the hypothesis of the *existence of "cofinally" many inaccessible cardinals* need not detain us, since that hypothesis is known to be equivalent to, and for present purposes may simply be taken to *be*, the hypothesis that every set is an element of some natural model of ZFC^2. The hypothesis of the *existence of a hyperinaccessible cardinal* may be taken to be the hypothesis that there exists a natural model of ZFC^2 + (cofinally many inaccessibles). The cardinal number of such a set is called a *hyperinaccessible* cardinal. All these hypotheses can be formulated in the language of ZFC—the second-order variables of ZFC^2 are not needed simply to talk about *models* of second-order theories.

Inaccessibles, hyperinaccessibles, superhyperinaccessibles, and so on, are only the first and smallest of large cardinals. There are many larger large cardinals—such as the so-called *indescribable* cardinals, of which there is a hierarchy of kinds—even among the so-called *small* large cardinals, those whose existence is consistent with $V = L$ (assuming ZFC plus the assumption of the existence of such a cardinal is consistent at all). One second-order set theory whose strength occupies a position in the hierarchy of indescribables, *Bernays set theory* B, will play a special role towards the end of this monograph, but I will defer description of it until then.

Beyond these are the large large cardinals, themselves divided into the medium large large cardinals (for each of which some modified version of $V = L$ has been developed, with which the existence of the cardinal in question is consistent if it is consistent at all), and the extra large large cardinals. The strongest large cardinal hypothesis of all, the existence of a *Reinhardt cardinal*, has been shown to be inconsistent; all the rest are presumed consistent by set theorists, though with decreasing confidence as one goes up the scale. Each new hypothesis implies new theorems expressible in the language of first-order arithmetic, and indeed theorems expressible by Π^0_1 formulas.[33]

The scale from Robinson to Reinhardt is summarized in table E, wherein $\approx$ indicates mutual interpretability or equal consistency strength. We will encounter systems at the levels of Q or $Q_2 = I\Delta_0$, of $Q_3 = I\Delta_0(\exp)$, of $F_4 = \Pi^1_1\text{-}CA_0$, of $P_2 = PA^2$, of $P^3 = PA^3$, and of B. Evidently, some modified Fregean systems are more limited than others as to how much of classical mathematics can be developed within them.

1.6 Philosophical Targets

Frege originally developed his system with the aim of showing that arithmetic and analysis are analytic, an aim thwarted when the system was found to be inconsistent. Neo-Fregeanism in a broad sense is the logical project of developing consistent modifications of Frege's inconsistent system. Neo-Fregeanism in a narrower sense is the logico-philosophical project of developing such systems *with an aim resembling Frege's*, which is to say, with the aim of establishing that a substantial amount of mathematics has some special epistemological status.[34] The two chapters that follow survey a diverse range of consistent modifications of Frege's system, some of which have been studied for decades and some of which appear here for the first time. In the case of each system, the logical question of how substantial a part of mathematics can be developed within it will be answered, at least in the sense that the (exact or approximate) place of the system in the fundamental series described in the preceding section will be indicated.[35]

That leaves the philosophical question of how special an epistemological status can be claimed for a given approach. Philosophical issues are notoriously far more difficult to settle than logical issues, and in the case of the older approaches, debate over philosophical issues has already gone on at great length in the literature, without any resolution being achieved or even approached.[36] The present work aims merely to characterize, not to resolve, philosophical issues. The present section is devoted to remarks on the general character of the philosophical debate. In the case of the older approaches, the remarks in the present section will be supplemented during the course of the survey of the diverse approaches by references to the philosophical literature where issues specific to a given approach have been debated. For ease of exposition, I will mainly restrict the discussion here to treatments of arithmetic.

A first generally applicable remark is that modified Fregean strategies typically have two aspects, just as Frege's original strategy did. On the one hand, there is a theorem to the effect that there exist certain objects and certain operations on them that satisfy the usual laws of arithmetic. In Frege's strategy, this is the theorem that certain abstracts with respect to equinumerosity, with appropriate operations on them, satisfy those laws. On the other hand, there is a definition identifying the natural numbers with the objects in question. In Frege's strategy, this is the definition of natural numbers as certain abstracts with respect to equinumerosity. If a special epistemological status is to be claimed for arithmetic, then such a status must be claimed both for the axioms, including the existence axioms, from which the existence theorem is deduced, and for the definition of natural number. Thus any version or variant of philosophical neo-Fregeanism will raise two different kinds of issues: first, over the philosophical status of the existence axioms; second, over the philosophical status of the definition identifying natural numbers with certain specified objects.

In connection with each of these two issues, philosophical neo-Fregeans find themselves confronted by rivals schools of philosophers of mathematics with rival programs of their own, but in neither case does one have to be a member of the rival school in question to share its main worries about philosophical neo-Fregeanism. On the existence question, philosophical neo-Fregeans are opposed by a largish number of *nominalists*, who have been convinced, by their reading of Benacerraf (1973) or otherwise, that we cannot have *any* kind of knowledge of non-concrete objects. But one need not be a nominalist to have doubts about the possibility of *analytic* knowledge of the existence of mathematical objects, or for that matter, of *any* objects. Many philosophers of otherwise diverse opinions share a conviction we cannot have such knowledge. Such philosophers look on philosophical neo-Fregeans the way skeptical spectators look on a stage magician who by shuffling cards seems to make an endless series of handkerchiefs appear. They are sure it must be trickery, even if they cannot quite say how the trick is worked. They are convinced that no amount of shuffing of our ideas can make a single real thing, let alone an endless series of real things, appear out of nothing.

The first task for philosophical neo-Fregeans is to reply to this general suspicion. And they do have available to them a general response to such general skepticism about the possibility of analytic knowledge of existence. It runs roughly as follows. The skeptics (this line of response suggests) are confusing two quite different kinds of situation that need to be distinguished. On the one hand, there is the kind of situation where we already have the notion of a certain sort of object, and principles about the existence and identity of such objects, and introduce a new term for those objects of that sort that happen to enjoy certain special properties. In such a situation, it is always in order to ask whether there *are* any objects of the given sort that enjoy those special properties, for the new term to apply to. On the other hand, there is the kind of situation where we are introducing a new term for objects of a new or previously unrecognized sort. In such a case, the meaning of the new term, the new notion it expresses, may be *constituted* by stipulations about the existence and identity of objects of the new sort. Apart from these stipulations, existence questions about the new objects have no meaning, while given the stipulations, some existence questions may have an immediate affirmative answer.[37]

To give a hackneyed but plausible example, if we have a geometrical theory with principles about points and lines and so forth, but do not yet have the notion of midpoint of a segment, someone may propose to introduce that notion by defining the *midpoint* of a line segment AC to be the unique point B with certain special properties, namely, those of lying on the segment AC, and of being such that the segments AB and BC are congruent. In that case, we can meaningfully ask whether the existence and uniqueness presupposition of this definition holds. The question is whether for every segment AC there exists a unique point B on the segment such that the segments AB and BC are congruent, which is something we can ask without using the term "midpoint". But the notion of direction (this line of response suggests) is *constituted* by the principle that every line has a direction, with two lines having the same one just in case they are parallel. If we do not already have the notion, and someone proposes to introduce it by stipulating these principles, it is a mistake to think we must first ask the question whether directions exist, and answer that question in the affirmative, before we accept these stipulations. On the contrary, before we accept the stipulations that give them meaning, the term "direction" and the question "Do directions exist?" are meaningless. After we accept the stipulations needed to make it meaningful, the question admits (given that we are already assuming the existence of lines) the immediate affirmative answer, "Yes, because every line has a direction."

So much for the first stage of the debate. The important point about it is that the general line of response to a general line of criticism that has just been sketched gives only a weak conclusion, precisely because it is so general, and applies not only to the plausible Fregean example of directions, but also to the unrestricted Fregean idea of extensions. The analyticity of "___s exist" is established only in the weak sense of "analytic" according to which we must accept the existence of ___s if we accept the notion of a ___. But analyticity in this sense this leaves entirely open whether we *should* accept the notion of a ___, and so leaves open whether we should accept that ___s exist. Acceptance of directions would be harmless, since adding the above-indicated principles about them to an existing theory is conservative.[38] Acceptance of extensions would be harmful, since as we have seen the unrestricted Fregean notion of them is inconsistent. The second task for philosophical neo-Fregeanism will be to argue why *its* preferred substitute for Frege's assumption—which substitute will typically be demonstrably non-conservative, but presumably consistent—has special virtues that make it more like the assumption of directions than like the assumption of extensions.

This second stage of debate will inevitably turn on the particular details of the strategy favored by the particular philosophical neo-Fregeans in question, and so we have come about as far as we can with the existence question while remaining at the high level of generality I have been maintaining in the present section.[39] One last general remark on the existence question may, however, be in order. It is just this, that if *most* of the work motivating a particular approach has to take place at the second stage of debate, and has to be based on claims to further virtues beyond analyticity in the weak sense described above, then it would seem that a neo-Fregean philosophical approach may not be all that *much* different from a non-Fregean philosophical approach that seeks to motivate its preferred foundations for mathematics—perhaps axiomatic set theory, perhaps something else—simply by appeal to such further virtues, without putting weight on analyticity in the weak sense.

Turning from the existence to the identity question, the question how to define the term "natural number," the main point that can be made about it at the present high level of generality is that the word "definition" is ambiguous. For there are at least three kinds of definitions in mathematics. Least problematic are *introductory* definitions, prescribing how a term not heretofore used in mathematics is to be used henceforth. Introductory definitions range from Hamilton's definition of "quaternions" as ordered quadruples of real numbers, to the definitions one finds in the work of Cantor. Cantor, notably, borrowed from grammar and introduced into mathematics the terms "cardinal" and "ordinal," essentially for abstracts with respect to equinumerosity of sets and with respect to isomorphism of well-orders, at the same time defining operations of cardinal and ordinal addition and multiplication and exponentiation.

One may indeed question the existential presuppositions of such a definition. One may, for instance, be as dissatisfied with Cantor's attempt to prove the existence of abstracts, by appeal to the supposed human power to create them through the exercise of a special mental faculty, as one is with Frege's attempt to prove the same conclusion, by appeal to the supposed logical law that all concepts have extensions. But I have done for the present with the existence question, and accepting the existence of the relevant objects, at least for the sake of argument, there can be no serious objection to taking some heretofore unused word and henceforth using it as a label for them. Granted that real numbers and ordered quadruples exist, there can be no serious objection to calling such of the latter as have all their components taken from among the former "quaternions." Granted abstracts with respect to the two relevant equivalences exist, there can be no serious objection to calling abstracts with respect to the one "cardinals" and abstracts with respect to the other "ordinals."

When we turn to definitions for terms already in use—and, of course, "number" is a term that has been in use since time immemorial—we must distinguish two further kinds of definition, which I will call *hermeneutic* and *revolutionary*. The former purport to describe what the term is at present used to mean, in the mathematical or in the general community, while the latter presume to prescribe what the term should be used to mean in future. My terminology "hermeneutic definition" and "revolutionary definition" may be novel,[40] but the things themselves should be familiar. Every entry in a mathematical dictionary or glossary of technical terms is a hermeneutic definition, intended to capture existing usage in the mathematical community. And there were many instances of revolutionary definitions during the great age of the instillation of rigor into modern mathematics, the age during which Cantor and Frege both flourished.

A good example from early in the era of rigorization would be Hamilton's redefinition of complex numbers as ordered pairs of real numbers. No one pretends or pretended that this is a correct hermeneutic definition, faithful to what mathematicians during the three centuries from Cardano to Hamilton had meant all along by "complex number." The aim of the definition, like that of other revolutionary redefinitions of the period, was to *replace* the previous notion by an improved one that would retain everything useful in the old one while eliminating anything obscure—beginning with the obscure idea that i and $-i$ are somehow "imaginary" in a sense in which 1 and -1 are not. A more immediately relevant example, from late in the era of rigorization, is provided by set-theoretic definitions of natural number, which again no one pretends or pretended are faithful to what mathematicians in the millennia since the Egyptians and Babylonians had meant all along.

In the case of any definition of "natural number," it is in order to ask whether the definition is supposed to be hermeneutic or revolutionary, and the question whether Frege's definition, or some present-day neo-Fregean's definition, is intended as hermeneutic or as revolutionary, is vital. On the one hand, a philosophical position is from many points of view much more interesting if its definitions are put forward as hermeneutic rather than revolutionary. For only if the definition is hermeneutic is the philosopher telling us anything about the actual and historical status of mathematics as it is and has been: if the definition is revolutionary, the philosopher is only telling us something about the the potential status of a proposed reformed mathematics.

But on the other hand, a hermeneutic definition is harder to establish than a revolutionary one. Or at least, definitions of the two kinds have to be evaluated in quite different ways. What one needs before adopting a hermeneutic definition is evidence of *truth*, in the sense of its fidelity to existing usage, whereas what one needs before adopting a revolutionary definition is evidence of *utility*. And crucially, when the evidence of truth is equally balanced between two incompatible descriptions, one has a powerful reason to abstain from believing either, whereas when evidence of utility is equally balanced between two incompatible prescriptions, though one is obliged to acknowledge that any choice of one over the other is arbitrary and conventional, nonetheless it remains perfectly appropriate to make an arbitrary and conventional choice. For two prescriptions may both be useful, even though only one can be used, whereas two incompatible descriptions cannot both be true.

And very often we do have equally balanced incompatible definitions. For instance, though Hamilton defined the complex number $a + bi = bi + a$ to be the ordered pair $\langle a, b\rangle$, it would have suited all his purposes equally well to define it to be $\langle b, a\rangle$. But since the definition was intended as revolutionary, it only follows that Hamilton's choice was not *uniquely* right, not that it was not *all right*. Similarly, though Zermelo gave a set-theoretic definition of natural number on which $2 = \{1\} = \{\{\emptyset\}\}$, and von Neumann gave one on which $2 = \{0, 1\} = \{\emptyset, \{\emptyset\}\}$, all the purposes of either of the two would have been served equally well by the other's definition. But again because these (grossly artificial) definitions were not intended as hermeneutic, if we share those purposes, then we may adopt either one of the pair of contrasting definitions as an arbitrary convention. By contrast, in either example, if the two definitions had been intended as hermeneutic, the existence of a rival with no obvious advantages or disadvantages would have been a serious objection against accepting either candidate.

On the identity issue, philosophical neo-Fregeans are opposed by a largish number of *structuralists*, who have been convinced, by the discussion of the Zermelo/von Neumann case in Benacerraf (1965) or otherwise, that no definition identifying natural numbers with objects of any sort not given as natural numbers can be hermeneutically correct (according to some, because the natural numbers are objects *sui generis*; according to others, because statements about "the" natural numbers are really generalizations about *all* systems of objects with the right properties). But one need not be a structuralist to be an opponent of any particular hermeneutic definition of natural number—for one reason, because it is enough to be a proponent of some rival hermeneutic definition.

This issue can be illustrated already in the case of Frege's original definition of natural number, after presenting a little backgound. To begin with, as Cantor noted, well-orders on equinumerous finite sets are isomorphic—this being a sophisticated counterpart of the homely fact that if you count the elements of a finite set, the result you get does not depend on the order in which you count them—so that there is a correspondence between finite cardinals and finite ordinals. Moreover, cardinal sums and products and powers correspond to ordinal sums and products and powers under this correspondence, so that in the finite case the cardinal and the ordinal versions of addition and multiplication and exponentiation obey exactly the same laws. Now it is a common mathematical practice, called "abuse of language," to omit to distinguish notationally between items that are distinct notionally, when the distinction makes no difference to the laws obeyed. Thus, though the positive integer +2, the rational number 2/1, and the real number 2.000 … are all distinct notionally, mathematicians generally do not distinguish them notationally, as I have just done, except when forced to do so by symbolic computation programs like Mathematica™. As an instance of this phenomenon, owing to the correspondence between finite cardinals and ordinals, Cantor used the same notations, the same numerals 0, 1, 2, and so forth, for both. This is, of course, the same notation that is ordinarily used for the natural numbers, and moreover the common laws of finite cardinal and finite ordinal addition and multiplication and exponentiation are the same as those traditionally assumed for the natural numbers. All this raises the question of the relationship between natural numbers on the one hand, and finite cardinals and finite ordinals on the other.

Cantor himself was not greatly concerned with questions of this kind. The view of natural numbers that would most naturally fit with his other views would perhaps be one according to which the notion of natural number is an undifferentiated one, combining cardinal and ordinal aspects (which separate from each other only when we move on to the transfinite). Perhaps the product of *m* and *n* may be characterized, cardinal-style, as the number of items in an *m*-by-*n* array, but may equally be characterized, recursive-style, as the result of adding *m* to itself *n* times, without either of these characterizations or any characterization being "the" definition of multiplication.[41]

Frege, by contrast, quite definitely took natural numbers to be finite *cardinals*, not finite *ordinals*.[42] The important point for present purposes in all this is that one need not be a structuralist to find the existence of an ordinalist rival to Frege's cardinalist definition of natural number a serious problem for Frege's definition—*if* it is intended as hermeneutic.[43] Similarly, any arbitrary choice between two candidates would be a serious problem for any definition, *if* intended as hermeneutic, as more obviously would be the use of any conspicuously artificial device, such as the Weiner-Kuratowski treatment of ordered pairs.

There are some indications that Frege's definition *was* intended to be hermeneutic. For one thing, he does occasionally appeal to ordinary usage when criticizing the theories of number offered by other philosophers. For another thing, he represents himself as disagreeing with Kant over the nature of numbers and our knowledge of them, and unless he meant by "number"—and for that matter, by "analytic"—what Kant meant by the same term, it is hard to see how Frege could be disagreeing with him, rather than just talking past him. There is nothing in Frege quite corresponding to the striking passage in Russell's *Introduction to Mathematical Philosophy* where he virtually confesses that his own logicist definition of number must be regarded as a *re*definition, saying, "It is . . . more prudent to content ourselves with the class of couples, which we are sure of, than to hunt for a problematical number 2 which must always remain elusive."[44]

In evaluating a program, one must attend not only to how much mathematics one gets, and not only to what axioms, and especially what existence axioms, one needs to get it, but also (if there is any chance the program is intended as a hermeneutic rather than a revolutionary one) to the character of its definitions. Do they involve any arbitrary choices or artificial devices? Do they ignore anything that seems to be an ingredient of the intuitive notion (as any approach to arithmetic that is too exclusively cardinalist or ordinalist may be suspected of doing); or do they, inversely, include anything that does not seem to be part of the intuitive notion (as any approach to numbers that identifies them with objects not given as numbers may be suspected of doing)?[45]

I will therefore aim, in the survey that follows, to make clear for each strategy considered what *all* its relevant features are: its scope and limits, its axiomatic basis, and the character of its definitions. After having done my best to make those features clear, I will for the most part leave the ultimate philosophical evaluation of the strategy to the reader, thus making the present work primarily a contribution to neo-Fregeanism in the broad rather than the narrow sense.

2

Predicative Theories

RUSSELL'S WORK first drew attention to the notion of predicativity,[1] though the predicativity restriction he imposed in his theory of types was essentially undone by his axiom of reducibility. This chapter is devoted to modified Fregean systems that impose, and do not subsequently undo, serious predicativity restrictions on the assumption of the existence of concepts, but that, unlike Russell's system, freely allow the existence of extensions for those concepts allowed to exist.

Michael Dummett, the most prolific and in many ways the most influential figure in Frege studies in the latter half of the last century, expressed in Dummett (1991) the minority view that the real source of the problem in Frege's system is the admission of impredicative comprehension for concepts. Partial support for Dummett's position is supplied by a result of Terence Parsons (1987), who showed that dropping comprehension does render Frege's system consistent, as had been conjectured by Peter Schroeder-Heister (1987). But George Boolos (1993) argued that this fact in itself is hardly sufficient to justify Dummett's claim, unless it can be shown that dropping or restricting comprehension does not interfere too badly with the project of deriving the basic laws of arithmetic within the system.

In effect, Dummett's claim presents us with a challenge to determine just how much mathematics can be done in a predicative Fregean system. This challenge was first taken up by Richard Heck (1996), who found that Q could be interpreted in a predicative version of Frege's system. An alternative development of a somewhat larger fragment of arithmetic in a somewhat weaker Fregean system will be presented here, deriving from joint work of the author with A. P. Hazen.[2] Other, related published work will also be surveyed in the present chapter, along with some new results.

2.1 A Simple Predicative Theory

Let us begin with the simplest imaginable kind of predicative version of Frege's system. We have two styles of variables, x, y, z, … for objects, and X, Y, Z, … for predicative concepts, we have the identity symbol $=$ for objects, and the extension symbol $\ddagger$, which applies to a variable of type X to form a term $\ddagger X$ that can be substituted for variables of type x. There are just two axioms, *predicative* comprehension for one-place concepts only, and Law V.

(1) $\exists X \forall x(Xx \leftrightarrow \phi(x))$

provided $\phi(x)$ *contains no bound concept variables*

(2) $\ddagger X = \ddagger Y \leftrightarrow \forall z(Xz \leftrightarrow Yz)$

Let me call this system PV (with "P" for "predicative" and "V" pronounced "five"). We may then introduce the notation $\{x\colon \phi(x)\} = \ddagger$«$x\colon \phi(x)$» for the set of x such that $\phi(x)$, when such a set exists, and define sethood and elementhood in the usual Fregean way:

(3) $\text{ß}y \leftrightarrow \exists Y(y = \ddagger Y)$

(4) $x \in y \leftrightarrow \exists Y(y = \ddagger Y \;\&\; Yx)$

A point needs to be made here about these abbreviations, however. What the proviso in (1) means is that $\phi(x)$ is to contain no bound concept variables *when written out in primitive notation*, unpacking all abbreviations. Can $\phi(x)$ contain expressions of the form $\{x: \psi(x)\}$? Let us see what happens as we unpack the definition of, say, a singleton, as it occurs in a formula:

(5) $\phi(\{z: z = x\})$

(5a) $\exists y(y = \ddagger\text{«}z: z = x\text{»} \;\&\; \phi(y))$

(5b) $\exists y \exists Z(\forall z(Zz \leftrightarrow z = x) \;\&\; y = \ddagger Z \;\&\; \phi(y))$

Not to put too fine a point on it, (5) eventually unpacks into something that involves the forbidden bound concept variables. So the answer to our original question is: no, set terms $\{x: \psi(x)\}$ may *not* appear in $\phi(x)$ in (1).

A *rank-zero* set term is one of the form $\{x: \phi(x)\}$ where no set terms appear in ϕ. For instance, $\{x: x = u\}$, the term for the singleton of u, is such a term. A *rank-one* set term is one of the form $\{x: \phi(x)\}$ where rank-zero set terms may appear in $\phi(x)$, but no others. For instance, $\{x: \exists u(x = \{y: y = u\})\}$, the term for the set of singletons, is such a term. Terms of rank two and higher are similarly defined. In PV, rank-zero set terms can be introduced as abbreviations, and the following version of Law V proved for them:

(6) $\{x: \phi(x)\} = \{x: \psi(x)\} \leftrightarrow \forall x(\phi(x) \leftrightarrow \psi(x))$

But what the observations of the preceding paragraph show is that *rank-one* set terms are in general not available in PV, because a formula $\phi(x)$ with rank-zero set terms involves implicit quantification over concepts, and the comprehension axiom of PV does not apply to give us a concept «$x: \phi(x)$» and thence a concept extension or set $\ddagger$«$x: \phi(x)$» $= \{x: \phi(x)\}$.

It goes without saying that we get extensionality when sets are introduced as extensions. We also get the various existence axioms listed in table F. For instance, the null set we obtain as $\{y: y \neq y\}$. Adjunction we obtain as $\{y: Uy \vee y = v\}$ where $u = \ddagger U$. The rest will be left to the reader.

EXACTLY HOW MUCH SET THEORY we get in this way can be described using the *Löwenheim-Behmann* theorem. This theorem is often quoted simply as saying that there is an effective procedure to decide whether a given formula of monadic first-order logic with identity—first-order logic with no function symbols and all relation symbols one-place except for the identity symbol—is logically valid. This much is due to Löwenheim (1915). However, the proof of the theorem in Behmann (1922) actually establishes more.[3]

Let $\boldsymbol{u}$ represent a finite string of variables $u_0, u_1, \cdots, u_r$ and similarly $\mathbf{P}$ a finite string of one-place relation symbols $\mathrm{P}_1, \mathrm{P}_2, \cdots, \mathrm{P}_s$. The theorem says that if $\phi(\boldsymbol{u}, \mathbf{P})$ is a formula of monadic first-order logic with no relation symbols but the P_i and no free variables but the u_i and with n bound variables, then ϕ is equivalent to a disjunction of conjunctions, where each conjunction is of the following special kind. First, each of its conjuncts includes for each u_i and u_j exactly one of $u_i = u_j$ or $u_i \neq u_j$, Second, each includes for each u_i and P_j exactly one of $\mathrm{P}_j u_i$ or $\sim \mathrm{P}_j u_i$. Third, each includes for each of the 2^s possible combinations

$$(7) \qquad (\sim)\mathrm{P}_1 v \ \& \ldots \& \ (\sim)\mathrm{P}_s v$$

where the negation symbol may be present or absent in front of each of the s relation symbols, a statement about the number of v distinct from all the u_i for which this combination holds. This statement either says that there are exactly k such v for some $k < n$, or that there are at least n such v.[4] Here the formula $\exists_k v\ \psi(v)$ expressing that there are at least k objects v such that $\psi(x)$ looks like this in the cases $k = 2$ and $k = 3$:

$$\exists v_1 \exists v_2(\psi(v_1) \ \& \ \psi(v_2) \ \& \ v_1 \neq v_2)$$
$$\exists v_1 \exists v_2 \exists v_3(\psi(v_1) \ \& \ \psi(v_2) \ \& \ \psi(v_3) \ \& \ v_1 \neq v_2 \ \& \ v_1 \neq v_3 \ \& \ v_2 \neq v_3)$$

The formula $\exists_k!v\ \psi(v)$ expressing that there are exactly k objects v such that $\psi(v)$ may be taken to be $\exists_k v\ \psi(v)\ \&\ {\sim}\exists_{k+1}v\ \psi(v)$.[5] Conjunctions involving contradictions of either of the types $x = y\ \&\ y = z\ \&\ x \neq z$ or $x = y\ \&\ Px\ \&\ {\sim}Py$ may be dropped. If the original formula was contradictory, all conjunctions will be dropped and we will have an empty disjunction, which by convention we equate with logical falsehood.

We can apply this result to a second-order formula $\phi(x, \boldsymbol{u}, \boldsymbol{W})$ with no bound second-order variables, and consider what it tells us about what we get as $\{x: \phi(x, \boldsymbol{u}, \boldsymbol{W})\}$ for given values of the parameters $\boldsymbol{u}$ and $\boldsymbol{W}$. The theorem tells us that ϕ is equivalent to a compound obtained by disjunction, conjunction, and negation from formulas not mentioning x, formulas of type $x = v$ or $v = x$, where v is u_i or $\ddagger W_j$, and formulas of type $W_j x$. Disjunction, conjunction, and negation correspond to union, intersection, and complement. The set $\{x: \psi\}$, where ψ does not mention x, is Ø or V according as ψ is false or true. The set $\{x: x = v\}$ or $\{x: v = x\}$ is the singleton $\{v\}$. The set $\{x: W_j x\}$ is $\ddagger W_j$. Since intersection is obtainable from complement and union, every set we get for whatever ϕ and whatever values of the parameters is obtainable from the null set, singletons $\{v\}$ of object parameters and of extensions $\ddagger W$ of concept parameters, and those extensions themselves, by union and complementation. All the set-existence axioms we get are implied by null set, singleton, union, and complement. A similar analysis shows that if we considered an *ultra*predicative theory UPV, where concept parameters were not allowed in (1), the set theory we would get could be axiomatized with null set, singleton, pair, triple, . . . , and complement.

Names for several systems in this area are listed in table G. PF, for *predicative Fregean* set theory, and UPF, for *ultrapredicative Fregean set theory*, are the sets theories obtained from PV and UPV as just described. ST, for *Szmielew-Tarski* *set theory*, is a subsystem of PF (adjunction being a consequence of union and singletons), and UST a system related to UPF as ST is related to PF. The still weaker system UUST is of interest because if we adopt Zermelo's definition of 0 as $\varnothing$ and x' as $\{x\}$, then in UUST we can prove the Peano postulates other than induction, or what is the same thing, the first two axioms (Q1), (Q2) of Q and Q*. Inversely, if we have (Q1) and (Q2) we get an interpretation of UUST by relativizing quantifiers to the formula $x = 0 \vee \exists y(x = y')$ and defining $x \in y$ to abbreviate $x' = y$. Adding monadic or dyadic predicative second-order logic is indicated by prefixing P or P_2, respectively, to the name of the theory, though the difference between monadic and dyadic makes no difference for any system for which we have pairing, since dyadic R can be simulated by monadic X using ordered pairs (which we get in the usual way from unordered pairs). Later we will consider the result of adding Zermelo's axiom of separation to ST, giving STZ.

In the next section I will prove the following:

> ***Extended Szmielew-Tarski Theorem.*** Robinson arithmetic is interpretable in each of the three set theories ST and PUST and P_2UUST.

The result for ST is due to Wanda Szmielew and Tarski, and is mentioned without proof in the monograph of Tarski, Mostowski, and Robinson (1953). A proof was published subsequently by Collins and Halpern (1970). Montagna and Mancini (1994) obtained a slight strengthening by a different method, and as often happens in such situations, the new proof is from some points of view more interesting than the new result. The strengthening consists in showing that extensionality is not needed, which is significant from some perspectives, but not from the Fregean. The method is an adaptation of the Solovay-Nelson method of proving other theories interpretable in Q to prove Q interpretable in other theories. The method I use in the next section is a simplification, formulated so as to cover the cases of the other two theories mentioned.

The common features of the three cases are first, that they all give (Q1) and (Q2), and second, that each in one way or another makes available *relations* R, S, T, ... subject to the following existence axioms:

(R1) $\exists R \forall x \forall y \sim Rxy$

(R2) $\forall R \forall u \forall v \exists S \forall x \forall y (Sxy \leftrightarrow Rxy \vee (x = u \;\&\; y = v))$

We may call the R in (R1) the *empty* relation E, and the S in (R2) the *adjunction* R_{uv} of u and v (in that order) to R. It would be convenient to have, in addition to the two-place, relations some one-place relations X, Y, Z,... with the analogous properties of nullity and adjunction:

$$\exists X \forall x \sim Xx$$
$$\forall X \forall u \exists Y \forall x (Yx \leftrightarrow Xx \vee x = u)$$

And in fact we do have them, in effect, if we have (R1) and (R2). For we can interpret $\exists X$ as $\exists R$ if we interpret Xx as Rxx. (In effect, X is represented by the identity relation on X.)

As to how we get (R1) and (R2), that differs in the three cases. The pairing axiom gives ordered pairs $\langle u, v\rangle = \{\{u\}, \{u, v\}\}$. In ST, where we have null set and adjunction, we can take the relations R to be sets of ordered pairs. In PUST we cannot, since we lack adjunction, but we can take the relations R to be predicative concepts under which ordered pairs fall. In P_2UUST, we have no ordered pairs, but can take the relations R to be predicative relational concepts.

2.2 Interpreting Arithmetic

What will be shown in this section is that in the two-sorted theory with axioms (Q1), (Q2), (R1), (R2) there is, for suitable definitions of $<$ and $+$ and $\cdot$, a formula ϕ such that each of the axioms of Q* holds for all objects for which ϕ holds, and ϕ is multiplicative in the sense defined in the discussion of Q* in the preceding chapter. The first task will be to define $<$. To that end, call X *regressive* if Xx holds whenever Xx' does, and define $x < z$ to hold if and only if, for any X, if X is regressive and such that Xy holds if $y' = z$, then Xx holds. We quickly obtain the following:

(Q1*) $\quad \sim x < 0$

(Q2*) $\quad x < z' \leftrightarrow x < z \vee x = z$

For (Q1*) consider the empty X. This X vacuously fulfills the condition for being regressive, and also vacuously fulfills the condition that Xy for y with $y' = 0$, since there are no such y by (Q1). But $\sim Xx$ and therefore $\sim x < 0$. For the left-to-right direction of (Q2*), suppose $x < z'$ and $x \neq z$, and let X be regressive, and suppose Xy holds if $y' = z$. Adjoin z to X, obtaining Y. This Y fulfills the condition for being regressive, since the only new item is $z = y'$ and we have Xy by assumption, and there is no other u with $z = u'$ by (Q2). Y also fulfills the condition that Yu holds if $u' = z$, since x does. Since $x < z'$, we have Yx, and since $x \neq z$, the only new item, we have Xx, showing that $x < z$. For the right-to-left direction of (Q2*), suppose first $x = z$, and let X be regressive and such that Xy holds if $y' = z'$. Since $x = z$, $x' = z'$ and so Xx holds, showing $x < z'$. Suppose now instead $x < z$, and let X be regressive and such that Xy holds if $y' = z'$, so Xz holds. Then if $u' = z$, we have Xu by definition of regressive, so since $x < z$ we have Xx, showing that $x < z'$.

The reasoning here is largely Frege's.

Let us use the usual abbreviation

(1) $$x \leq y \leftrightarrow x < y \vee x = y$$

Then (Q1*) and (Q2*) immediately yield the following:

(2) $$x \leq 0 \leftrightarrow x = 0$$

(3) $$x \leq z' \leftrightarrow x \leq z \vee x = z'$$

Consider now the inductive proof of transitivity reviewed in the discussion of systems of arithmetic in the preceding chapter. What this proof proves, from the present perspective, is that the following formula $\phi_0(x)$ is inductive:

(4) $$\forall y \forall z(z < y \;\&\; y < x \rightarrow z < x)$$

Now let $\phi(x)$ be any inductive formula. It is easily seen that the conjunction of two inductive formulas is inductive (and similarly for orderly, additive, and multiplicative). So if we let $\psi(x)$ be the conjunction of $\phi(x)$ and $\phi_0(x)$, then $\psi(x)$ is inductive. Let us use in the usual way the bounded quantifiers $\forall y < x$ and $\forall y \leq x$ and $\exists y < x$ and $\exists y \leq x$. Now let $\phi^<(x)$ be the formula $\forall z \leq x\ \psi(z)$. Trivially, $\phi^<(x)$ implies $\psi(x)$ and hence implies both $\phi(x)$ and $\phi_0(x)$.

I make three further claims. I first claim $\phi^<(0)$ holds. Indeed, $\phi^<(0)$ is equivalent by (2) to $\psi(0)$, which holds since ψ is inductive. I second claim that if $\phi^<(x)$ holds, then $\phi^<(x')$ holds, making $\phi^<$ inductive. Indeed $\phi^<(x')$ is equivalent by (3) to $\phi^<(x)$ & $\psi(x')$, and we are assuming $\phi^<(x)$, which implies $\psi(x)$, which in turn implies $\psi(x')$ since ψ is inductive. I third claim that if $\phi^<(x)$ holds and $y < x$, then $\phi^<(y)$ holds, making $\phi^<$ orderly. Indeed, since we are assuming $\phi^<(x)$, which implies $\phi_0(x)$, (4) above is applicable. From (4) it follows that for all $z \leq y$, we have $z \leq x$, and therefore have $\psi(z)$ by $\phi^<(x)$. This suffices to show that $\phi^<(y)$ holds and to complete the proof.

The orderly formula $\phi^{<}(x)$ will be called the *orderly kernel* of ϕ. Let $\phi_1 = \phi_0{}^{<}$. Then ϕ_1 implies ϕ_0, ϕ_1 is orderly, and besides (Q1), (Q2), (Q1*), (Q2*), the auxiliary law of transitivity also holds for all objects for which ϕ_1 holds.

Now consider the next axiom:

(Q3*) $$x = 0 \vee \exists y(y < x \;\&\; x = y')$$

Obviously (Q3*) holds for $x = 0$. Even *without* assuming (Q3*) holds for x, (Q3*) holds for x', taking x as y, since $x < x'$ by (Q2*). So (Q3*) is inductive. Hence so is its conjunction ψ with ϕ_1. If we let $\phi_2 = \psi^{<}$, then ϕ_2 implies ϕ_1, ϕ_2 is orderly, and all the axioms of Q* down through (Q3*) hold for objects for which ϕ_2 holds.

We may call such objects *orderly*. Note that, whereas by (Q3*) every orderly object is either zero or a successor, by (Q1) no object is both: the zero and successor cases are mutually exclusive as well as jointly exhaustive. For the remainder of the proof, we restrict our attention to orderly objects: the English "all objects" is to be understood as elliptical for "all orderly objects," the formal $\forall x(\ldots)$ is to be understood as an abbreviation for $\forall x(\phi_2(x) \rightarrow \ldots)$.

We are now ready to tackle addition. For a given object y, define a relation R to be a *computation of sums with* y if we have the following:

(5a) $$\exists v R0v$$

(5b) $$\forall v(R0v \rightarrow v = y)$$

(6a) $$\forall u(\exists w Ru'w \rightarrow \exists v Ruv)$$

(6b) $$\forall u \forall w \forall v(Ru'w \;\&\; Ruv \rightarrow w = v')$$

Define R to be a *computation of a sum of* y *with* x *if further* $\exists v Rxv$. Define R to be a *computation of a sum of y with x as v* if further Rxv. Define w to be *a* sum of y with x, and write Σyxw, if there exists a computation of a sum of y with x as w. If $\exists! w \Sigma yxw$, we call this w the *sum* $y + x$ of y and x; otherwise the sum $y + x$ is undefined. Call x *summable* if $\forall y \exists ! w \Sigma yxw$, or in other words, if for all y the sum $y + x$ is defined, and further this sum is what it ought to be, by which I mean that we have the following:

(7a) $\quad x = 0 \rightarrow y + x = y$

(7b) $\quad x = w' \rightarrow y + w$ is defined & $y + x = (y + w)'$

If we had mathematical induction available, we could prove, by a more or less standard set-theoretic argument, that every x is summable, using the arithmetical results we already have. I will set out the proof in somewhat tedious detail so as to make plain what assumptions are used. Let x be any object, arbitrary but fixed throughout the proof.

For $x = 0$, we have by (R1) an empty relation E and by (R2) the result $R = E_{0y}$ of adjoining 0 and y in that order to it, so that we have Ruv if and only if $u = 0$ and $v = y$. (5a) and (5b) follow immediately, and (6a) and (6b) hold vacuously, since the zero and successor cases are exclusive. So R is a computation of a sum with y of 0 as y, and we have $\Sigma y0y$. Moreover, we cannot have $\Sigma y0w$ for any $w \neq y$ by clause (5b) in the definition of computation. So y is the unique w such that $\Sigma y0w$, and $y + 0 = y$, in agreement with (7a). (7b) holds vacuously. So zero is summable.

For x', first of all note that by (Q2) we do not also have $x' = u'$ for any $u \neq x$: the predecessor x of x' is unique. Assuming x is summable, we may prove x' is summable as follows. Let $y + x = w$. Then there is a computation S of the sum with y of x as w. We have by (R1) the result $R = S_{x'w'}$ of adjoining x' and w' in that order to S, so that we have Ruv if and only if Suv or $u = x'$ and $v = w'$. And by the exclusiveness of the zero and successor cases, and the uniqueness of predecessors, in the one new case $u = x'$ and $v = w'$, the first component x' is not zero and is not the successor of anything but x. This noted, and recalling that (5) and (6) hold for S, it easily follows that (5) and (6) hold for R, which therefore is a computation of a sum with y of x' as w', and we have $\Sigma yx'w'$. Moreover, we cannot have $\Sigma yx'v$ for any $v \neq w'$, because if T is a computation of a sum with y of x' as v, then by clause (6a) in the definition of computation, T is also a computation of a sum with x of y as something u, which can only be w by the summability of x. Clause (6b) then requires that $v = u' = w'$. So w' is the unique v such that $\Sigma yx'v$, and $y + x' = w'$, in agreement with (7b). (7a) holds vacuously. So x' is summable.

The reasoning here is largely Dedekind's.

In the present context, what this argument shows is that the conjunction ψ of ϕ_2 with the formula expressing that x is summable is inductive. If we let $\phi_3 = \psi^<$, then ϕ_3 implies ϕ_2, ϕ_3 is orderly, and all the axioms of Q* down through (Q4) and (Q5) hold for all objects for which ϕ_3 holds, (Q4) and (Q5) being immediate from (7a) and (7b). Moreover, the objects for which ϕ_3 holds are all summable.

Now for a crucial point: we are *not* in a position to assert that ϕ_3 is additive. For this would require not just that the sum $y + x$ *exists* for all y and x for which ϕ_3 holds, but also that the sum *satisfies* ϕ_3. Before we can get additivity, we will need to establish two auxiliary laws, the law that addition *preserves order*, and the law that addition is *associative*:

(8) $y \leq x \rightarrow z + y \leq z + x$

(9) $z + (y + x)$ is defined & $z + (y + x) = (z + y) + x$

Note that if $\phi_3(x)$ and $y \leq x$, then $\phi_3(y)$ and therefore not just x but y also is summable, and the two sums mentioned in (8) exist. Note also that if $\phi_3(x)$ and $\phi_3(y)$ and $\phi_3(z)$, so that all three are summable, the sums $(y + x)$ and $(z + y)$ and $(z + y) + x$ are all defined. We are not, however, in a position to assert that $(y + x)$, or for that matter either of the other sums just mentioned, is summable, or that $z + (y + x)$ is defined. That it is so has to be explicitly stated in (9) as part of what is to be proved.

As in so many previous cases, it would not be hard to prove (8) and (9) using mathematical induction. For (8) the base step is trivial, and at the induction step, assuming (8) for x and considering any $y \leq x'$, we either have $y \leq x$, in which case $z + y \leq z + x$ by induction hypothesis and therefore $z + y \leq (z + x)' = z + x'$, or else $y = x'$ and $z + y = z + x'$, and we are done. For (9), the proof for the base step consists of a series of equations derived from (Q4), while that for the induction step consists of a series of equations derived from (Q5) and the induction hypothesis, thus:

(10a) $(z + y) + 0 = z + y = z + (y + 0)$

(10b) $(z + y) + x' = ((z + y) + x)' = (z + (y + x))'$
$= z + (y + x)' = z + (y + x')$

In the by-now usual manner, these proofs can be adapted to give us an orderly ϕ_4 implying ϕ_3 and an orderly ϕ_5 implying ϕ_4 such that for objects satisfying ϕ_5 we have all the laws derived so far down through (8) and (9).

Now for any inductive formula $\phi(x)$, let $\psi(x)$ be the conjunction of $\phi^<(x)$ with ϕ_5, and let $\phi^+(x)$ be the formula

$$\psi(x)\ \&\ \forall z(\psi(z) \rightarrow \psi(z + x))$$

Trivially, $\phi^+(x)$ implies $\psi(x)$ and hence implies both $\phi(x)$ and $\phi_5(x)$.

I make four further claims. First, I claim that $\phi^+(0)$ holds. Indeed, the first conjunct $\psi(0)$ holds, since ψ is inductive, and the second conjunct is immediate, since for all z we have $z + 0 = z$ by (Q4).

Second, I claim that if $\phi^+(x)$ holds, then so does $\phi^+(x')$, making ϕ^+ inductive. The first conjunct $\psi(x')$ follows, since ψ is inductive, from $\psi(x)$, which in turn follows from $\phi^+(x)$. For the second conjunct, suppose that $\psi(z)$ and consider $z + x'$. We have $\psi(z + x)$ by $\phi^+(x)$, and then $\psi((z + x)')$, since ψ is inductive, and so we have $\psi(z + x')$, since $z + x' = (z + x)'$ by (Q5).

Third, I claim that if $\phi^+(x)$ holds and $y < x$, then $\phi^+(y)$ holds, making ϕ^+ orderly. The first conjunct $\psi(y)$ follows, since ψ is orderly, from $\psi(x)$, which in turn follows from $\phi^+(x)$. For the second conjunct, suppose that $\psi(z)$ and consider $z + y$. We have $\psi(z + x)$ by $\phi^+(x)$, and we have $\psi(w)$ for any $w \leq z + x$, since ψ is orderly, and so we have $\psi(z + y)$, since $z + y \leq z + x$ by (8), which we have, since $\phi^+(x)$ implies $\psi(x)$ which implies $\phi_5(x)$ which implies $\phi_4(x)$.

Fourth, I claim that if $\phi^+(x)$ and $\phi^+(y)$ hold, then $\phi^+(y + x)$ holds, making ϕ^+ additive. The first conjunct $\psi(y + x)$ follows from $\phi^+(x)$ and $\psi(y)$, which latter follows in turn from $\phi^+(y)$. For the second conjunct, suppose that $\psi(z)$ and consider $z + (y + x)$. We have $\phi(z + y)$ from $\phi^+(y)$ and then $\phi((z + y) + x)$ from $\phi^+(x)$, and so we have $\phi(z + (y + x))$, since $z + (y + x) = (z + y) + x$ by (9), which we have since $\phi^+(x)$ implies $\phi_5(x)$.

The additive formula $\phi^+(x)$ will be called the *additive kernel* of ϕ. Let $\phi_6 = \phi_5{}^+$. Then ϕ_6 implies ϕ_5, ϕ_6 is additive, and (Q0)–(Q5) hold for all objects for which ϕ_6 holds. We may call such objects *additive*. For the remainder of the proof, we restrict our attention to additive objects.

The treatment of multiplication will be given only in outline. We begin by defining a notion Πyxw of there being a computation of the product with y of x as w, and when $\exists! w\ \Pi yxw$, defining that w to be the product $y \cdot x$. We then call x *productive* if for any y the product $y \cdot x$ is defined. A more or less standard set-theoretic argument shows that the formula expressing that x is productive is inductive, and taking its additive kernel we obtain an additive ϕ_7 implying ϕ_6 such that all the axioms of Q* hold for objects satisfying ϕ_7.

We are *not* in a position to assert ϕ_7 is multiplicative. Before we can get multiplicativity, we will need to establish three auxiliary laws, the law that multiplication *preserves order*, the law that multiplication is *distributive* over addition, and the law that multiplication is *associative*:

(11) $$y \leq x \rightarrow z \cdot y \leq z \cdot x$$

(12) $$z \cdot (y + x) = z \cdot y + z \cdot x$$

(13) $$z \cdot (y \cdot x) = (z \cdot y) \cdot x$$

As in so many previous cases, these would not be hard to prove using mathematical induction. In the usual way—but now taking additive rather than just orderly kernels—we get successive additive formulas ϕ_8 implying ϕ_7, ϕ_9 implying ϕ_8, and ϕ_{10} implying ϕ_9, such that for objects satisfying ϕ_{10} we have all the laws derived so far, down through (11)–(13).

Now for any inductive formula $\phi(x)$, let $\psi(x)$ be the conjunction of $\phi^+(x)$ with $\phi_{10}(x)$, and let $\phi^{\cdot}(x)$ be the formula

$$\psi(x) \ \& \ \forall z(\psi(z) \rightarrow \psi(z \cdot x))$$

Trivially, $\phi^{\cdot}(x)$ implies $\phi(x)$. I make five further claims.

First, second, and third, I claim that $\phi^{\cdot}(0)$ holds, that $\phi^{\cdot}(x')$ holds if $\phi^{\cdot}(x)$ holds, and that if $\phi^{\cdot}(x)$ holds and $y < x$, then $\phi^{\cdot}(y)$ holds, making $\phi^{\cdot}$ inductive and orderly. The proofs are just like those used in the additive case, modulo replacement of addition by multiplication, and appeals to (Q4), (Q5), and (8) by appeals to (Q6), (Q7), and (11).

Fourth, I claim that if $\phi^{\cdot}(x)$ and $\phi^{\cdot}(y)$ hold, then $\phi^{\cdot}(y + x)$ holds, making $\phi^{\cdot}$ additive. Leaving the first conjunct to the reader, as to the second conjunct the key step is just this: assuming $\phi(z)$ and $\phi^{\cdot}(y)$ and $\phi^{\cdot}(x)$, the first and second give $\psi(z \cdot y)$, while the first and third give $\psi(z \cdot x)$. We then get $\psi(z \cdot y + z \cdot x)$, since ψ is additive, and hence $\psi(z \cdot (y + x))$ by (12), which we have, since $\phi^+(x)$ implies $\phi_9(x)$.

Fifth, I claim that if $\phi^{\cdot}(x)$ and $\phi^{\cdot}(y)$ hold, then $\phi^{\cdot}(y \cdot x)$ holds, making $\phi^{\cdot}$ multiplicative. Leaving the first conjunct to the reader, as to the second conjunct the key step is just this: assuming $\phi(z)$ and $\phi^{\cdot}(y)$ and $\phi^{\cdot}(x)$, the first and second give $\phi^{\cdot}(z \cdot y)$, and the third then gives $\phi^{\cdot}((z \cdot y) \cdot x)$, and hence $\phi^{\cdot}(z \cdot (x \cdot y))$ by (13), which we have, since $\phi^{+}(x)$ implies $\phi_{10}(x)$.

The formula $\phi^{\cdot}$ will be called the *multiplicative kernel* of ϕ Let $\phi_{11} = \phi_{10}^{\cdot}$. Then ϕ_{11} is multiplicative and all the axioms of Q* hold for all objects for which ϕ_{11} holds, and we can restrict our attention to multiplicative objects, defined in the obvious way, and we are done with the proof.

Understanding of the proof just concluded can be enhanced by considering what extensions we can and cannot get by the same methods. For we can by the same methods get interpretations of some extensions of Q* with more axioms, though not of others. It should be clear that for any law χ that could be proved by *quantifier-free* induction, we get an inductive formula whose multiplicative kernel will give us an interpretation of Q* + χ. A simple example would be any of the series of laws that lead up to the commutative law for addition, beginning with the first in the series:

(14) $$0 + x = x$$

The formula (14) is easy to prove inductive, and its multiplicative kernel is a multiplicative formula $\phi(x)$ such that all axioms of Q* + (14) hold for objects for which $\phi(x)$ holds, and this amounts to saying that all axioms of Q* + (14) hold when quantifiers are relativized to $\phi(x)$, which is what is needed to get an interpretation.

It is *not* in general the case that one can get an interpretation of Q* + χ for a law χ that has to be proved by Σ_1-induction, which is to say, induction on a condition involving an existential quantifier. This is not the case in general because of the gap between saying an existential formula holds for all objects for which $\phi(x)$ holds and saying that the existential formula holds with quantifiers relativized to $\phi(x)$. The one merely requires that something exist appropriately related to x, the other that this something itself satisfy ϕ. The gaps between summability and additivity, and between productivity and multiplicativity, were instances.

However, there are *some* cases of laws χ with existential quantifiers where we *can* get an interpretation of $Q^* + \chi$ by the methods we have been using. A simple example would be the *parity* law:

$$\exists y \leq x\,(x = y \cdot 2 \vee x = (y \cdot 2)') \tag{15}$$

The formula is not too hard to prove inductive, giving a multiplicative $\phi(x)$ such that all axioms of Q^* + (15) hold for objects for which $\phi(x)$ holds. Now to be sure, there *is* a gap between saying this and saying that all axioms of Q^* + (15) hold when quantifiers are relativized to $\phi(x)$, which is what is needed to get an interpretation. It is the gap between the following:

(16a) $\forall x(\phi(x) \rightarrow \exists y \leq x\,(x = y \cdot 2 \vee x = (y{\cdot}2)'))$

(16b) $\forall x(\phi(x) \rightarrow \exists y \leq x(\phi(y)\ \&\ (x = y{\cdot}2 \vee x = (y{\cdot}2)')))$

But *this* gap is bridgeable, *because the quantifier is bounded.* (16b) actually follows from (16a) and the *orderliness* of $\phi(x)$, which nearly enough is what is expressed by the following formula:

$$\forall x(\phi(x) \rightarrow \forall y \leq x\,\phi(y)) \tag{17}$$

By (17), the insertion of $\phi(x)$ in (16a) to produce (16b) is *redundant.*

The gap is, in fact, always bridgeable in the case of Δ_0-induction, which is to say, induction on a formula in which all quantifiers (if any) are bounded. This is owing to what is called the *absoluteness* of such formulas. Absoluteness means that the formula with quantifiers relativized to any suitable formula

$\phi(x)$ is equivalent to the formula with quantifiers unrelativized. If the bounds are variables, as in (17), "suitable" may be understood simply as orderly. If the bounds are variables and/or terms built up using $+$, suitability may be understood as additivity. If the bounds are variables and/or terms built up using $+$ and/or $\cdot$, suitability may be understood as multiplicativity.

For instance, suppose we have something of the following form:

(18a) $\exists y(y \leq x + x \;\&\; \forall z\,(z \leq u \cdot u \rightarrow \ldots))$

(18b) $\exists y(\phi(y) \;\&\; y \leq x + x \;\&\; \forall z\,(\phi(z) \;\&\; z \leq u \cdot u \rightarrow \ldots)$

If ϕ is multiplicative, then for x and u for which ϕ holds, (18a) and (18b) will be equivalent. For $\phi(x)$ implies $\phi(x + x)$, which with $y \leq x + x$ implies $\phi(y)$, while $\phi(u)$ implies $\phi(u \cdot u)$, which with $z \leq u \cdot u$ implies $\phi(z)$. And this makes the insertion of $\phi(y)$ and $\phi(z)$ redundant.

Using the same method over and over, we can get an interpretation of $Q^* + \chi_1 + \ldots + \chi_n$ for any laws χ_i that can be proved by quantifier-free or more generally Δ_0-induction. Such is the proof of Nelson's result that $I\Delta_0$ is "locally interpretable" in a theory once Q is interpretable in that theory. We do, however, when using this method, have to change the interpretation each time we want a new law. Something more than this is true, namely, Wilkie's result that $I\Delta_0$ is "globally interpretable" in a theory once Q is interpretable in that theory. But Wilkie's proof is beyond the scope of the present work.

One thing we can*not* get is exponentiation. It will be instructive to consider how the attempt to get it by the methods used for multiplication and addition breaks down. We can begin by defining a notion $Yyxw$ of there being a computation of the power of y to the x as w, and when $\exists! w Yyxw$ defining that w to be the power $y \uparrow x$. We can then call x *powerful* if for any y the power $y \uparrow x$ is defined. A more or less standard set-theoretic argument shows that the formula expressing that x is powerful is inductive, and taking its multiplicative kernel we obtain a multiplicative formula such that all the axioms of Q* hold for objects for which the formula holds, as besides do the following further axioms for exponentiation:

(Q8) $$x \uparrow 0 = 0'$$

(Q9) $$x \uparrow y' = (x \uparrow y) \cdot x$$

We are not, however, in a position to say that $x \uparrow y$ is powerful whenever x and y are, any more than we were in a position to say that $x \cdot y$ is productive when x and y are, or that $x + y$ is summable when x and y are. It is more or less clear by analogy what auxiliary laws we would want, to be able to go on and get a formula closed under exponentiation in the way that multiplicative formulas are closed under multiplication or additive formulas closed under addition. We would want a law that exponentiation preserves order, and laws stating how exponentiation distributes over addition and over multiplication, and finally, an associative law of exponentiation. And there is no difficulty over the first three of these:

(19) $$w \leq x \rightarrow y \uparrow w \leq y \uparrow x$$

(20) $$z \uparrow (y + x) = (z \uparrow y) \cdot (z \uparrow x)$$

(21) $$z \uparrow (y \cdot x) = (z \uparrow y) \uparrow x$$

But exponentiation simply *isn't* associative!

Suppose we ignore this fact and go on to define an *exponential kernel* $\phi^{\uparrow}$ using a definition of the following form:

$$\psi(x) \ \& \ \forall y(\psi(y) \rightarrow \psi(y \uparrow x))$$

Then we will want to make six claims, whose content the reader can guess. The first five of these we will be able to prove, using (Q8), (Q9), and (19)–(21). The last we will be unable to prove because the associative law is missing. All we will get is a multiplicative formula $\phi^{\uparrow}$ implying ϕ such that whenever $\phi^{\uparrow}(x)$ and $\phi^{\uparrow}(y)$, we have $\phi(x \uparrow y)$—but not necessarily $\phi^{\uparrow}(x \uparrow y)$.

2.3 Miscellany

In this section I will consider a variety of systems in the general neighborhood of the simplest predicative system PV: its double predicative companion P^2V; an un-Fregean system that "cheats" by helping itself to an intuitive notion of finitude; a predicative system PHP based on HP rather than Law V; and some curious second-order set theories in which the sets are not obtained from concepts as their extensions, but merely have concepts "floating" over them.

2.3a *A Double Predicative Theory*

Let us turn first to the second simplest kind of predicative Fregean system P^2V. We write the first-round concept variables $X, Y, Z, \ldots$ of PV now as $X^0, Y^0, Z^0, \ldots$ and the function symbol $\ddagger$ as $\ddagger^0$. We add second-round concept variables $X^1, Y^1, Z^1, \ldots$ and a function symbol $\ddagger^1$ taking them as arguments, with the following additional versions of comprehension and Law V:

(1) $\exists X^1 \forall x(X^1 x \leftrightarrow \phi(x))$

provided $\phi(x)$ contains no bound second-round variables

(2a) $\ddagger^1 X^1 = \ddagger^0 Y^0 \leftrightarrow \forall z(X^1 z \leftrightarrow Y^0 z)$

(2b) $\ddagger^1 X^1 = \ddagger^1 Y^1 \leftrightarrow \forall z(X^1 z \leftrightarrow Y^1 z)$

It will be recalled that in PV we had rank-zero set terms such as $\{u\}$ but no rank-one set terms such as $\{x\colon \exists u(x = \{u\})\}$, on account of the rank-zero terms involving implicit quantification over first-round concepts, making the comprehension axiom of PV inapplicable to them. The comprehension axiom of P^2V, by contrast, *is* applicable to them, so in P^2V we do get rank-one set terms, with the following version of (2) for them:

$$(2') \qquad \{x\colon \phi(x)\} = \{x\colon \psi(x)\} \leftrightarrow \forall x(\phi(x) \leftrightarrow \psi(x))$$

However, we do not get rank-two set terms in P^2V, because formulas involving rank-one set terms involve implicit quantification over *second*-round concepts.

It turns out that allowing a second round of predicative concepts can get us over the impasse we reached in trying to get exponentiation. In fact, for this purpose there is no need to use (2) for second-round concepts at all, and no need to do anything more with the first-round concepts beyond getting the set theory ST from them. The theory PST consisting of ST plus monadic predicative second-order logic is interpretable in P^2V, using the first-round concepts just to get ST and then forgetting about them, and making no use of $\ddagger^1$. And I claim Q_3 is interpretable in PST.

If we have monadic predicative second-order logic, the whole development of the preceding section can be redone for concepts in place of formulas. A concept is called *inductive* if 0 falls under it and x' falls under it whenever x does, and the notions *orderly* and *additive* and *multiplicative* are similarly defined. Then the proofs already outlined show that every inductive concept X^1 contains a multiplicative concept $Y^1 = X^{1\uparrow}$ such that the objects falling under Y^1 satisfy all the axioms (Q1)–(Q9) and whenever Y^1u and Y^1v, then $X^1(u \uparrow v)$. The one thing we can*not* claim is that Y^1 is *exponential* in the sense that besides being multiplicative Y^1 has the property that if Y^1u and Y^1v, then $Y^1(u \uparrow v)$.

However, we are in a position now to make one definition that we could not make previously, when we had only ST and first-order logic. We call an object x a *natural number* if it falls under every inductive concept X^1. This notion is expressible by a formula $\nu(x)$ but the formula is obviously an impredicative one, involving quantification over concepts X^1, and so does not determine a concept X^1. For purposes of finding an interpretation of the theory with axioms (Q1)–(Q9), this does not matter. I claim that the formula $\nu(x)$ is exponential.

The verification that 0 is a natural number, and that if x is a natural number, then so is x', will be left to the reader. As for orderliness, let x be a natural number and $y < x$. We want to show that y is a natural number. What we need to do is consider any inductive X^1 and show that y falls under X^1. To this end consider any orderly Y^1 contained in X^1. (We could have $Y^1 = X^{1<}$ or $Y^1 = X^{1\uparrow}$, for instance.) Since x is a natural number and Y^1 is inductive, x falls under Y^1, and since Y^1 is orderly and $y < x$, y falls under Y^1, and since Y^1 is contained in X^1, y falls under X^1 as required. Additive and multiplicative closure will be left to the reader. As for exponential closure, let u and v be natural numbers. We want to show that $u \uparrow v$ (is defined and) is a natural number. What we need to do is consider any inductive X^1 and show that $u \uparrow v$ falls under X^1. To this end consider $Y^1 = X^{1\uparrow}$. Since u and v are natural numbers and Y^1 is inductive, u and v both fall under Y^1. But then by definition of $Y^1 = X^{1\uparrow}$ it follows that $u \uparrow v$ falls under X^1, as required to complete the proof.

THE THEORY Q_3 ADDS Δ_0-induction to (Q1)–(Q9). Hazen observed (in a slightly different context) that allowing a second round of predicative concepts, as we have done, permits a comparatively easy derivation of Δ_0-induction—a *very* easy derivation compared to Wilkie's proof.

Let us first get clear what we are trying to prove. A Δ_0-induction axiom in general looks like this:

(3) $$\forall u(\phi(0, u) \ \&\ \forall x(\phi(x, u) \rightarrow \phi(x', u)) \rightarrow \forall x\phi(x, u))$$

where ϕ is a Δ_0 formula. The formula ϕ is built up from equations and inequalities $t = s$ and $t < s$ between terms themselves built up from variables using 0 and $'$ and $+$ and $\cdot$ and $\uparrow$, using negation, conjunction, disjunction, conditional, biconditional, and *bounded* quantifications.

What (3) looks like when relativized is this:

(3′) $\forall u(\nu(u) \rightarrow (\phi^{\nu}(0, u)\ \&\ \forall x(\nu(x) \rightarrow (\phi^{\nu}(x, u) \rightarrow \phi^{\nu}(x', u))) \rightarrow$
$\forall x(\nu(x) \rightarrow \phi^{\nu}(x, u))))$

wherein the superscript $^{\nu}$ indicates relativization of the quantifiers in ϕ. The crucial observation is the *absoluteness* of Δ_0 formulas, already discussed in the preceding section, according to which we actually don't have to worry about relativizing the quantifiers in ϕ, because we have the following:

(4) $$\nu(y)\ \&\ \nu(z) \rightarrow (\phi^{\nu}(y, z) \leftrightarrow \phi(y, z))$$

Given the principle of absoluteness, what we need to prove is then not (3′) but just the following:

(5) $\forall u(\nu(u) \rightarrow (\phi(0, u)\ \&\ \forall x(\nu(x) \rightarrow (\phi(x, u) \rightarrow \phi(x', u))) \rightarrow$
$\forall x(\nu(x) \rightarrow \phi(x, u)))$

or reverting to the practice of suppressing parameters, the following:

(5′) $$\phi(0)\ \&\ \forall x(\nu(x) \rightarrow (\phi(x) \rightarrow \phi(x'))) \rightarrow \forall x(\nu(x) \rightarrow \phi(x))$$

So assume ϕ holds for zero and for the successor of any natural number for which it holds, and let x be any natural number. We need to prove ϕ holds for x. To this end, let $\psi(w)$ be the following formula:

(6) $$w \leq x \rightarrow \phi(w)$$

It will suffice to prove that $\psi(w)$ holds for all natural numbers w, and for this it will suffice to show that $\psi(w)$ is an inductive formula, since this formula $\psi(w)$—unlike, for instance, the formula $\nu(x) \rightarrow \phi(x)$, which involves the notion of natural number—involves no quantification over second-round concepts, and therefore determines a second-round concept, which will be an inductive concept if the formula is an inductive formula, and which in that case will have every natural number falling under it.

Now $\psi(0)$ holds, since we are assuming $\phi(0)$. So suppose $\psi(w)$ holds, to prove $\psi(w')$ holds. There is nothing to prove unless $w' \leq x$, in which case $w \leq x$ and is therefore a natural number, and moreover a natural number for which $\phi(w)$ holds, since $\psi(w)$ does. But then $\phi(w')$ holds and hence $\psi(w')$ holds, since we are assuming ϕ holds for the successor of any natural number for which it holds. This completes the proof.

As to *what* the proof proves, it may be stated as follows, writing PST for monadic predicative second-order logic with ST.

Doubly Extended Szmielew-Tarski Theorem. Kalmar arithmetic is interpretable in PST.

2.3b *An Un-Fregean Digression*

Let me now digress a moment. Logicians have sometimes considered enriching first-order logic with a new quantifier $\mathbf{F}x$ intended to mean "there exist only finitely many x". The following three principles seem evident on such an understanding of the notation:

(F1) $\sim\exists x\phi(x) \rightarrow \mathbf{F}x\,\phi(x)$

(F2) $\mathbf{F}x\,\phi(x) \rightarrow \mathbf{F}x(\phi(x) \vee x = y)$

(F3) $\mathbf{F}x\,\phi(x)\ \&\ \forall x(\psi(x) \rightarrow \phi(x)) \rightarrow \mathbf{F}x\,\psi(x)$

If one has no things, one has only finitely many things. If one has finitely many things and one more, one has only finitely many things. If one has some things from among finitely many things, one has only finitely many things.

If one wanted to develop a theory of (all and only) finite sets using the new quantifier, the obvious way to proceed would be to introduce symbols ß and $\in$ for set and element, and in addition to the axiom of extensionality, a single set-existence axiom (scheme), as follows:

$$(1) \qquad \exists y(\text{ß}y \mathbin{\&} \forall x(x \in y \leftrightarrow \phi(x))) \leftrightarrow \mathsf{F}x\, \phi(x)$$

The things for which ϕ holds form a set if and only if there are only finitely many of them.

Three set-existence axioms follow immediately, corresponding to the three laws (F1)–(F3). Two of these we have already seen, the null set and adjunction axioms of ST, corresponding to (F1) and (F2). The other is the axiom scheme of Zermelo's axiom of separation, in the following form:

$$(2) \qquad \forall y(\text{ß}y \rightarrow \exists z(\text{ß}z \mathbin{\&} \forall x(x \in z \leftrightarrow x \in y \mathbin{\&} \psi(x))))$$

Taking down the scaffolding of the new quantifier, let us just consider the first-order set theory STZ consisting of ST plus (2).

The next result is well-known to specialists.[6] There is, however, no generally agreed attribution for the result in question, which places it in the category called mathematical "folklore."

Folk Theorem. First-order Peano arithmetic is interpretable in STZ.

The machinery developed at the beginning of the proof of the theorem about the interpretability of Q permits a proof of this fact requiring little additional work. Let us think of that proof especially as it applies to ST, so the R and X are just certain sets. Follow the proof down to the introduction of the notion of *orderly kernel.*

Next we need some auxiliaries. First, I claim the formula $0 \leq x$ is inductive. The easy proof is left to the reader. Second, I claim the following formula is also inductive:

$$\exists u \forall y (y \in u \leftrightarrow y \leq x) \tag{3}$$

Note that if such a u exists it is unique by extensionality; we may then call it in words the *segment* up to x, and in symbols $[x]$. To show that $[x]$ exists, which is to say that (3) holds, for $x = 0$, note that by (Q1*) we have $[0] = \{0\}$. To show $[x']$ exists if $[x]$ exists, note that by (Q2*) we have $[x'] = [x] \wedge x'$.

Third, for a given x, we may say that a set u is *inductive up to x* if $0 \in u$ and $w' \in u$ whenever $w \in u$ and $w \leq x$. Call x *inductible* if $x \in u$ for every u that is inductive up to x. I claim the formula expressing inductibility is inductive. That 0 is inductible is immediate from the definition of "inductive up to x." Supposing x is inductible, it may be shown that x' is inductible as follows. Let u be any set that is inductive up to x'. We first show u is also inductive up to x. And indeed we have $0 \in u$, and moreover if $w \in u$ and $w' \leq x$, then by (Q2*) we have $w' \leq x'$, and so $w' \in u$, since u was inductive up to x'. Since u is thus inductive up to x and x is inductible, we have $x \in u$. But then since u is inductive up to x' and $x' \leq x'$, we have $x' \in u$, to complete the proof.

Let now $\delta(x)$ be the orderly kernel of the conjunction of the three auxiliary formulas just shown to be inductive, and call the objects for which $\delta(x)$ holds *natural numbers*. Since $\delta(x)$ is orderly, 0 is a natural number, x' is a natural number whenever x is, and w is a natural number whenever x is and $w \leq x$. Also, for any natural number x we have $0 \leq x$, $[x]$ exists, and x is inductible.

Now let $\phi(x)$ be any formula such that $\phi(0)$ holds, and such that if $\phi(x)$ holds and x is a natural number, then $\phi(x')$ holds. I claim that $\phi(x)$ holds for every natural number x. The reasoning is not dissimilar to that in the proof of Δ_0-induction in the preceding subsection. To see that $\phi(x)$ holds for every natural number x, let x be any natural number, and apply separation to $[x]$ to obtain a set u such that for all w we have:

$$w \in u \leftrightarrow w \leq x \ \& \ \phi(w) \tag{4}$$

I claim u is inductive up to x. We have $0 \in u$, since x is a natural number and therefore $0 \leq x$, and since $\phi(0)$ by our hypothesis about ϕ. Suppose we have $w \in u$ and $w' \leq x$. To prove $w' \in u$ first note that since $w \in u$, we have $\phi(w)$. Then note that since $w \leq w' \leq x$ and x is a natural number, so are w' and w. Therefore $\phi(w')$ holds by our hypothesis about ϕ, and hence $w' \in u$ as required to show u inductive up to x. Since x is inductible, we have $x \in u$, and therefore $\phi(x)$.

Thus the full, unrestricted principle of mathematical induction for all formulas of the language applies to natural numbers. What were in the proof of the interpretability of Q merely proofs that certain formulas are inductive now become proofs that certain laws hold for natural numbers. These laws include, besides the facts from which we started, that zero is not a successor and that the successor function is one-to-one, the fact that sums and products obeying the recursion equations (Q4)–(Q7) exist. But those laws, together with induction, are all the axioms of P^1.

The idea of somehow using an intuitive notion of finitude as a foundation for first-order arithmetic (and in particular, to evade the kind of limitations of predicative approaches that will be established in the next sections) turns up in several different guises in the recent literature on foundations of mathematics.[7] I feel it would be out of place, however, in the present monograph to survey this literature or elaborate further than I have already done. For it seems clear to me that no idea could be more non-, un-, and anti-Fregean than that of helping oneself to intuitions about finitude as axioms, not proved as theorems from logical axioms and a suitable definition of finitude.

2.3c *Predicativity and Hume's Principle*

Let us next peek ahead briefly at a subject that will be treated more fully in the next chapter. Modified Fregean systems that dispense with extensions and take their start from Hume's principle are primarily a topic for that chapter; however, it is natural to consider *predicative* versions or variants of such an approach here.

Actually, I will consider only the simplest imaginable approach of this kind. We have three styles of variables, x, y, z, … for objects, and X, Y, Z, … for first-round predicative concepts, henceforth just called "concepts," and R, S, T, … for first-round predicative two-place relational concepts, henceforth just called "relations." We have the identity symbol = for objects, and the number symbol #, which applies to any variable of type X to form a term $\#X$ that can be substituted for variables of type x.

There are just three axioms, *predicative* comprehension for concepts and relations, and Hume's principle, as follows:

(1) $\exists X \forall x(Xx \leftrightarrow \phi(x))$
where $\phi(x)$ contains no bound concept variables

(2) $\exists R \forall x \forall y(Rxy \leftrightarrow \psi(x, y))$
where $\psi(x, y)$ contains no bound concept variables

(3) $\#X = \#Y \leftrightarrow X \approx Y$

Let me call this simple predicative system with Hume's principle PHP, by analogy with PV. I claim that the two-sorted system with axioms (Q1), (Q2), (R1), (R2) can be interpreted in PHP, and hence that Q can be.

As a preliminary, several notions and notations pertaining to concepts and relations that will be used below are collected in tables H and I.

Note that the empty relation and adjunction for relations immediately give us axioms (R1) and (R2). So it only remains to give definitions of 0 and $'$ and find some formula $\phi(x)$ such that axioms (Q1) and (Q2) hold for all objects for which ϕ holds. (This amounts to interpreting UUST in PHP.) Much of the work has been done for us by Frege. We may follow him in defining zero and immediate succession, but must depart from him in defining order:

(4) $$0 = \#\varnothing$$

(5) $$\$xy \leftrightarrow \exists X \exists w(x = \#X \;\&\; {\sim}Xw \;\&\; y = \#(X \wedge w))$$

(6) $$x \lhd y \leftrightarrow \exists X \exists Y(x = \#X \;\&\; X \subset Y \;\&\; y = \#Y)$$

I write $\lhd$ for the auxiliary notion of order needed for the following construction, to leave open the connection of this auxiliary relation $\lhd$ to any order relation $<$ one may introduce later once we have found an interpretation of Q.

We can use Frege's proof for the following fact:

(7) $$\$xy \;\&\; \$xz \rightarrow y = z$$

The proof runs as follows. Given $x = \#X_1 = \#X_2$ and w_1 and w_2 with ${\sim}X_1w_1$ and ${\sim}X_2w_2$ and $y = \#(X_1 \wedge w_1)$ and $z = (X_2 \wedge w_2)$, then since $\#X_1 = \#X_2$, there is a one-to-one correspondence R such that $X_1 \approx_R X_2$, and what we need to show is that there is a one-to-one correspondence S such that $X_1 \wedge w_1 \approx_S X_2 \wedge w_2$, which will show $y = z$. Such an S is provided by $R \wedge \langle w_1, w_2 \rangle$. The details I am leaving to the reader here (and at corresponding places below) are those of the verification that S is indeed a one-to-one correspondence if R is.

By (7), if there exists a y such that $\$xy$, it is unique, and may be called *the* immediate successor x' of x. It is immediate from the definitions that $\sim\$0x$, and this fact and (7) give us (Q1) and (Q2) in the following qualified forms:

If x' exists, then	$0 \neq x'$
If x' and y' exist, then	$x' = y' \rightarrow x = y$

We need three lemmas about order. The first lemma says:

$$(8) \qquad x \lhd y \;\&\; y = \#Y \rightarrow \exists X(X \subset Y \;\&\; x = \#X)$$

The proof runs as follows. Since $x \lhd y$, there are $X_1 \subset Y_1$ with $x = \#X_1$ and $y = \#Y_1$. Since also $y = \#Y$, there is a one-to-one correspondence R such that $Y_1 \approx_R Y$. We want to show there is an $X \subset Y$ with $x = \#X$. It suffices to let X be the image $R[X_1]$. A one-to-one correspondence S such as is needed to show $x = \#X$ will be provided by the restriction $R \mid X_1$.

The second lemma says:

$$(9) \qquad x \lhd y \;\&\; y \lhd z \rightarrow x \lhd z$$

The proof runs as follows. Since $y \lhd z$, there exists $Y \subset Z$ with $y = \#Y$ and $z = \#Z$. But then by (8), since $x \lhd y$ there exists $X \subset Y$ with $x = \#X$. It follows that $X \subset Z$ and so $x \lhd z$.

The third lemma says:

$$(10) \qquad \$yz \rightarrow (x \lhd z \leftrightarrow x \lhd y \vee x = y)$$

The proof runs as follows. Assume there exist Y and w with $y = \#Y$ and $\sim Yw$, and $z = \#(Y \wedge w)$. For the left-to-right direction of the biconditional in (10), assume $x \lhd z$. Then by (8), there exists $X_1 \subset Y \wedge w$ with $x = \#X_1$. To show $x \leq y$, it will suffice to show there exists $X_2 \subset Y$ with $x \leq \#X_2$. If $\sim X_1 w$, we may take X_2 for X_1 and we are done. If $X_1 w$, there must be a v

with $\sim X_1 v$ but Yv, in which case we let $X_2 = (X_1 - w) \wedge v$. It then remains to show that $\#X_2 = x = \#X_1$, which is to say, that there is some one-to-one correspondence R such that $X_1 \approx_R X_2$. Such an R is provided by $I_{X\text{-}w} \wedge \langle w, v\rangle$. For the right-to-left direction, it is immediate from the definitions that $y = \#Y \lhd \#(Y \wedge w) = z$. If $x \leq y$, then $x \lhd z$ by (9).

It is immediate from the definitions that $\sim x \lhd 0$, and this fact and (10) give us (Q1*) and (Q2*), with the latter in qualified form:

$$\sim x \lhd 0$$

$$\textit{If } y' \textit{ exists, then} \qquad x \lhd y' \leftrightarrow x \lhd y \vee x = y$$

What remains to be done is to find a formula $\phi(x)$ such that, calling the objects for which it holds *protonatural numbers*, 0 will be a protonatural, and for all protonatural x, the successor x' will exist, *and will also be protonatural*. Then (Q1) and (Q2), without qualifications, will hold relativized to ϕ.

The definition is just this, that x is a *protonatural* if and only if the following holds:

$$(11) \qquad \exists X(\forall y(Xy \leftrightarrow y \lhd x) \;\&\; x = \#X \;\&\; \sim Xx))$$

It may be remarked that since the definition of $y \lhd x$ involves bound concept and relation quantifiers, we are not in a position to assert that for *every* x there will exist a concept X holding of all and only the $y \lhd x$. If there does, we may call it in words the *strict segment* of x and in symbols $\langle x\rangle$.

Now 0 is protonatural with Ø for the X in the definition (11). For $\langle 0\rangle = \text{Ø}$ by (Q1), $0 = \#\text{Ø}$ by definition of 0, and $\sim$Ø0 by definition of Ø. Now suppose x is protonatural, and let X be as in definition (11). Then $\#(X \wedge x)$ is the successor x' of x. I claim x' is protonatural with $X \wedge x$ for the X in the definition (11). By (10) we have $\langle x'\rangle = \langle x\rangle \wedge x = X \wedge x$, and hence the first as well as the second conjunct of the definition hold.

What remains to be proved is what is asserted in the third conjunct, namely, that x' does not fall under $X \wedge x = \langle x \rangle \wedge x$. That is to say, we have to exclude the possibility that $x' \lhd x$ and the possibility that $x' = x$. By (10) we have $x \lhd x'$. If $x' \lhd x$, then by (9) we have $x \lhd x$, while if $x' = x$, we have $x \lhd x$ directly. But $x \lhd x$ is precluded by the first and third conjuncts of (11) for x.

Thus we have proved the following:

> ***The Predicative Variant of Frege's Theorem.*** The axioms of Robinson arithmetic can be deduced in dyadic predicative second-order logic from Hume's principle and suitable definitions of arithmetical notions.

Adding a second round of predicative concepts (with no need for relations in the second round), to obtain a system that may be called P^2HP, would permit the interpretation of Q_3 in much the same manner as with P^2V.

2.3d *Zig-Zag Theories*

Before he settled on his theory of types as a solution to his paradox, Russell (1906) contemplated two other approaches that might be taken, the *limitation of size* and *zig-zag* theories. Limitation of size has always been a central idea in Cantorian set theory, and consideration of Cantorian set theory from a Frege-inspired point of view is postponed until the next chapter. As for the zig-zag theory, Russell described it as assuming that a "propositional function" has an extension if it is simple enough, but not if it is too complex. Russell had something specific in mind, an idea that he experimented with for a couple of years before committing himself to type theory, and that is now of purely historical interest.[8] But the general idea of a theory with full, impredicative comprehension for concepts, and some kind of simplicity restriction on which concepts are supposed to have extensions, may be briefly considered here.

One immediate difficulty with this idea is the following. A "propositional function," or sense of a predicate or open formula, can presumably be said to be simple or complex just as a formula can be. But a concept, or referent of a predicate or open formula, cannot in itself be called "simple" or "complex." For any two formulas that hold of the same objects determine the same concept, and a concept may be determined by several formulas of quite different degrees of simplicity and complexity. Presumably we have to understand a concept to be "simple enough" if there is *at least one* formula determining it that is "simple enough."

To implement this idea, let us go back to the framework for safe-extensions theories briefly considered in the preceding chapter. Thus we have two styles of variables, x, y, z, … for objects, and X, Y, Z, … for (one-place, first-level) concepts. We have full, impredicative comprehension for concepts. And writing €xX for "x is the extension of X", we have extensionality in the following Fregean form:

$$(1) \qquad €xX \;\&\; €yY \rightarrow (x = y \leftrightarrow X \equiv Y)$$

The assumption that X has an extension if it is simple may then be expressed thus:

$$(2) \qquad \forall y(Xy \leftrightarrow \phi(y)) \rightarrow \exists x\, €xX$$
provided $\phi(x)$ *is simple*

Sethood ß and elementhood $\in$ may then be defined in the usual Fregean way.

A little thought shows that "simple" will have to be so understood as to exclude the appearance of concept variables as parameters. Else the very simplest case, where $\phi(y)$ is Yy, Y being a parameter, would give the existence of an extension for every concept Y. If also "simple" is taken to imply *predicative*, then $\phi(y)$ will in fact have to be *first-order*. The result will amount to the first-order set theory UPF, but now with full, impredicative second-order logic.

An alternative approach would be to have, in addition to variables X, Y, Z, … for concepts in general, special variables X^0, Y^0, Z^0, … for "simple" ones, with a restricted version of comprehension for these, and only these allowed to have extensions. This is more flexible, in that it can accommodate "simple" concept parameters. Again if "simple" is understood as *predicative*, we will get something familiar, PF, but again with full, impredicative second-order logic.

We have now encountered several different systems consisting of a weak first-order set theory (PF, UPF, ST, UST, UUST) plus some variety of second-order logic. For UUST, where pairing is not available, one gets a very weak system unless one adds *dyadic* second-order logic; for the others, there is no difference between monadic and dyadic. Where the second-order logic is predicative, there may be slight differences in strength of the combined theory; at any rate, we got an interpretation of Q_3 with ST as the underlying set theory, and only Q_2 with UST, though these differences will be smoothed out by ramification, permitting more rounds of predicative concepts. With full, impredicative second-order logic, the differences among the various underlying weak set theories are obliterated. The weakest of them, combined with so powerful a logic, take us all the way up to second-order Peano arithmetic P^2 (and the strongest of them takes us no further). Let me state this result as a named theorem:

> ***Dedekind's Theorem, Version II.*** Second-order Peano arithmetic is interpretable in UUST with dyadic second-order logic and is also interpretable in UST with monadic second-order logic.

The proof is left as an exercise to the reader.[9]

There is something peculiar about all these systems, at least from the point of view of those familiar with more conventional Cantorian systems of set theory that admit concepts (under the label "classes") in addition to sets. In the latter one generally starts with a first-order set theory T involving some axiom *scheme*, like separation

(3) $$\exists u \forall w (w \in u \leftrightarrow w \in x \;\&\; \phi(w))$$

When one moves to a second-order version T', this scheme gets replaced by a single axiom:

(3′) $$\exists u \forall w (w \in u \leftrightarrow w \in x \;\&\; Xw)$$

The individual instances of (3) are recovered from (3′) and corresponding instances of comprehension:

(4) $$\exists X \forall x (Xx \leftrightarrow \phi(x))$$

In such a conventional system T', the "classes" interact with the sets through (3′) or other second-order axioms replacing what in the first-order version were schemes. The result of the interaction is that more can be proved in T' using "classes" than could be proved in T alone; by this I mean more can be proved *just about sets*. Questions that can be formulated in the language of T but not answered in T get answers in T': T' is *not* a conservative extension of T.

To be sure, if we consider only adding *predicative* comprehension, then the resulting theory—call it T°—is a conservative extension of T by *Shoenfield's theorem*, which says that one can always add predicative second-order logic to a first-order theory, and in the process if there are any schemes in the original theory, replace them by single axioms, and the result will be a conservative extension. But this is a somewhat surprising result, requiring a non-trivial proof.

By contrast with all this, in the case of the kinds of combinations of a set theory with a second-order logic that we have been considering, there is no scheme that gets replaced by a single axiom, or anything of the sort. The sets and "classes" or concepts do not interact: the concepts just "float" over the sets, so to speak. We may be able to interpret stronger theories in the second-order theory than in the first-order theory, but this is *not* because we can prove anything more about the objects of the first-order theory, but only because we can single out special ones among them by means of second-order formulas that we could not single out by means of first-order formulas.

Not only is it true for such theories with "floating" concepts that the *predicative* versions are conservative extensions—that much would follow from Shoenfield's theorem, and specifically from the degenerate case of Shoenfield's theorem where there are no schemes to be replaced—but so are the *impredicative* versions. The proof is very easy, assuming familiarity with basic ideas of model theory.

Now I *do* assume familiarity with such basic results of model theory as the soundness and Gödel completeness theorems and the Löwenheim-Skolem theorem, but let me remark that these hold for many-sorted first-order theories as well as for one-sorted first-order theories. To specify a model for a one-sorted theory, one has to specify a non-empty set, the *domain* over which its variables are to be thought of as ranging, and for each k-place relation symbol in the language, a k-place relation on that set, the *denotation* of the symbol, and similarly for function symbols. To specify a model for a two-sorted theory, one has to specify two domains, one for the one sort of variable and the other for the other. (It does not matter if the domains overlap and the same object appears in two capacities, so to speak.) When, say, a binary relation symbol takes variables of the first sort in the first place and variables of the second sort in the second place, the relation specified as its denotation must be between objects in the first domain and objects in the second. Apart from these obviously necessary changes, all model-theoretic notions work the same way in the many-sorted as in the one-sorted case.

Recall that, so far as questions about what is provable are concerned—and conservativeness is, of course, like consistency a question about what is provable—second-order theories are (or can be viewed as) simply many-sorted first-order theories. The only models with which we will be concerned are $\in$-*models*, in which the domain of the concept variables consists of certain subsets—perhaps all of them, perhaps not—of the domain of the object variables, and the denotation of the usually unwritten falling-under symbol ∇ is the relation of set to element. In other words, a "concept" of the model will be a set of "objects" of the model, and "object" will fall under a "concept" if and only if it is an element thereof. Similarly, "relational concepts" will be "relations," which may be construed as sets of ordered pairs. The axiom of extensionality will come out true in any $\in$-model.

The set A of those "objects" of the model that satisfy a formula $\phi(x)$ (for given values of its parameters), and *only* that set A, will satisfy $\forall x(Xx \leftrightarrow \phi(x))$. Hence the instance of comprehension for $\phi(x)$ will come out true in the model (for the given values of the parameters) if and only if that set A is included among the "concepts" of the model. This A will obviously be so if the domain of the concept variables, the "concepts" of the model, include *all* subsets of the domain of the object variables. An $\in$-model of this kind, where the domain of the X-variables is the full power set or set of all subsets of the domain of the object variables, is called a *standard* model, and is a model of full, impredicative comprehension.[10]

To return now to the conservativeness claim stated above, let T be a first-order theory. I first claim that if T is consistent, then the result T' of adding full, impredicative second-order logic to T is also consistent. For if T is consistent, by the

Gödel completeness theorem it has a model. As just explained, this model can be turned into a model of T'—a standard model. Since T' thus has a model, it is consistent as claimed. The conservativeness result follows immediately from the consistency result. For if ϕ is a formula of the language of T that is not provable in T, then $T + \sim\phi$ is consistent, and applying the consistency result to it, it follows that $T' + \sim\phi$ is consistent, and ϕ is not provable in T'.

The foregoing argument may first have been made by Hao Wang, who considered several examples of a theory T' consisting of full, impredicative monadic second-order logic "floating" over a first-order set theory T.[11] Indeed, we may call such theories as T' *Wang-style* set theories with at least as much justice as we can call them *zig-zag* theories.

The best-known Wang-style set theory is indeed one based on a considerably stronger underlying first-order set theory than any of the weak ones we have considered here. It is second-order set theory that Wang proposed as a replacement for the inconsistent system ML introduced in the first edition of Quine's *Mathematical Logic*. In this case, the first-order set theory is a system of Quine's called NF or "New Foundations." For detailed discussion of the systems NF and ML, which are beyond the scope of the present study, see Quine (1951).

NF has interested a number of writers because it allows for a universal set—in this respect it is a zig-zag rather than a limitation of size theory—though we have seen a couple of other, simpler set theories PF and UPF that do so as well, and will see another in the next chapter. The idea underlying the choice of existence axioms in NF is un-Fregean. Moreover, for NF the consistency of the system remains in serious doubt to this day. (It has not been proved consistent relative to any theory in whose consistency there is any very general confidence among specialists.) For these reasons, I will leave NF, and more generally Wang-style set theories and the zig-zag idea, aside here, and return to the proper subject of this chapter, *predicative* theories.

2.4 Consistency: Model-Theoretic Proofs

One might well hope that if one round of predicative concepts gets us Nelson-Wilkie arithmetic Q_2, with addition and multiplication, and if two rounds gets us Kalmar arithmetic Q_3, with exponentiation, then three rounds would get us Gentzen arithmetic Q_4, with superexponentiation, and that if we continued through the whole hierarchy of a ramified predicative theory, we would get all primitive recursive functions. Nothing of the sort is the case, for a reason (pointed out to the author in a different context by Saul Kripke) that is closely connected with the reason why I have attached Gentzen's name to Q_4.

The reason is simply this: that the *consistency* of the theories PV and P^2V and the analogous P^3V, P^4V, …, as well as that of the theories in the parallel series beginning with PHP, can be proved using "finitistic" proof-theoretic methods pioneered by Gentzen, which are known to be formalizable in Q_4. But Gödel's second incompleteness theorem says, roughly speaking, that no consistent theory T can prove its own consistency, and implies that no theory that is interpretable in T can prove the consistency of T. It follows that if the consistency of a theory T can be proved in Q_k, then the Q_m for $m \geq k$ cannot be interpreted in T. In this way we get upper bounds on how large a fragment of arithmetic can be interpreted in our predicative Fregean systems.

In this section I will describe "infinitistic" model-theoretic consistency proofs for the various predicative Fregean systems; in the next I will describe what needs to be done to convert these into "finitistic" proof-theoretic consistency proofs. My goal in this section will be to describe models for the second-order theories PV and P^2V, by a method that obviously can be extended to P^3V and beyond. As for PHP, it is a subtheory of full, impredicative second-order logic plus HP—a system often called *Frege arithmetic*—for which a simple model-theoretic consistency proof will be discussed in the next chapter.

THE CONSISTENCY OF PV and P^2V will be proved in this section by proving the consistency of certain associated impredicative theories, $\wp V$ and $\wp^2 V$, which I must now introduce. It is convenient to think of $\wp V$ as being obtained in two stages. First let L_1 be the language with object variables $x, y, z, \ldots$ and first-round concept variables $X^0, Y^0, Z^0, \ldots$, and just the identity symbol = for objects. Let T_1 be the theory whose axioms are just the instances of the full, impredicative comprehension scheme:

(1) $$\exists X^0 \forall x (X^0 x \leftrightarrow \phi(x))$$
where $\phi(x)$ *is a formula of* L_1

Note that $\phi(x)$ is permitted to contain bound first-round concept variables.

Now let L_2 add the extension symbol $\ddagger^0$ and let $T_2 = \wp V$ add Law V:

(2) $$\ddagger^0 X^0 = \ddagger^0 Y^0 \leftrightarrow \forall z (X^0 z \leftrightarrow Y^0 z)$$

Now (1) gives full, impredicative comprehension for formulas *not involving* $\ddagger^0$. Hazen points out that for formulas *without bound concept variables*, each instance of comprehension for a formula that *does* involve $\ddagger^0$ follows from an instance that does *not*. For if there are no bound concept variables, in $\phi(x, \boldsymbol{u}, \boldsymbol{W}^0)$, where $\boldsymbol{u}$ stands for a string of object parameters u_i and $\boldsymbol{W}^0$ for a string of concept parameters W_i^0, the only way $\ddagger^0$ can appear is attached to the W_i^0. Take more object variables v_i and replace each occurrence of $\ddagger^0 W_i^0$ by v_i to obtain a formula

$\psi(x, \boldsymbol{u}, \boldsymbol{v}, \boldsymbol{W}^0)$ from which $\phi(x, \boldsymbol{u}, \boldsymbol{W}^0)$ can be recovered by substituting $\ddagger^0 W_i^0$ for v_i. Then (1) tells us that $\psi(x, \boldsymbol{u}, \boldsymbol{v}, \boldsymbol{W}^0)$ determines a concept for *all* values of the parameters, and since this includes the values $v = \ddagger^0 W_i^0$, it follows that $\phi(x, \boldsymbol{u}, \boldsymbol{W}^0)$ determines a concept. Thus we have the following strengthened version of (1):

(1′) $\exists X^0 \forall x (X^0 x \leftrightarrow \phi(x))$
where $\phi(x)$ *is a formula of* L_2 *that*
either *contains no bound concept variables* $X^0, Y^0, Z^0, \dots$
or *does not contain the symbol* $\ddagger^0$

This strengthened version (1′) includes the comprehension scheme for PV, which allows no bound concept variables.

The instance of comprehension that got Frege into trouble, it will be recalled, was for $\exists X(x = \ddagger X \,\&\, {\sim}Xx)$. This troublesome instance involves *both* a bound concept variable *and* an occurrence of ‡. Insisting that the same formula may not involve both turns out to be enough to block the paradox.[12]

It is convenient to think of $\wp^2$V as being obtained in two more stages. First let L_3 add second-round concept variables $X^1, Y^1, Z^1, \dots$, and let T_3 add comprehension:

(3) $\exists X^1 \forall x (X^1 x \leftrightarrow \phi(x))$
where $\phi(x)$ *is a formula of* L_3

Now let L_4 add the extension symbol $\ddagger^1$ and let $T_4 = \wp^2$V add the following two new forms of Law V:

(4a) $\ddagger^1 X^1 = \ddagger^0 Y^0 \leftrightarrow \forall z (X^1 z \leftrightarrow Y^0 z)$

(4b) $\ddagger^1 X^1 = \ddagger^1 Y^1 \leftrightarrow \forall z (X^1 z \leftrightarrow Y^1 z)$

We get a strengthened version (3′) of (3) related to it as (1′) is related to (1), and this strengthened version (3′) includes the comprehension scheme for P^{2}V.

A model for $\wp^2$V will be obtained in four stages. First, let M be any countably infinite set, and take M as the domain for the object variables. If we take the full power set $\wp(M)$, or set of all subsets of M, as the domain for the first-round concept variables, we will get a model of the comprehension axioms (1). But there is no way with *this* model to assign a denotation to $\ddagger^0$ that will give us a model of $T_2 = \wp$V. For axiom (2) will hold if and only if the denotation of $\ddagger^0$ maps the domain of the first-round concept variables one-to-one into the domain of the object variables, and $\wp(M)$ cannot be mapped one-to-one into M by Cantor's theorem.

So instead we apply the Löwenheim-Skolem theorem, which applied to this situation tells us that there is a countable subset M^0 of $\wp(M)$ such that taking M^0 as the domain of the first-round concept variables *still* gives a model of T_1. Since M^0 and M are both countably infinite, there will be a one-to-one function from the former into the latter. With an eye to further developments, choose such a one-to-one function f^0 for which infinitely many elements of M are *not* in the range of f^0. (For instance, fix enumerations of M^0 and M and map the first, second, third, . . . elements of the former to the first, third, fifth, . . . elements of the latter.) Taking f^0 as the denotation of $\ddagger^0$ then will suffice to give us a model of $T_2 = \wp$V.[13]

We now repeat the process. To get a model of T_3, first take as the domain of the second-round concept variables the full power set $\wp(M)$. Of course this means that the sets in M^0 will be doing double duty, as possible values of first-round concept variables and as possible values of second-round concept variables; but that does not matter.

Now apply the Löwenheim-Skolem theorem again to get a countable subset M^1 of $\wp(M)$ that, taken as the domain of the second-round concept variables, still gives a model of T_3. Since there are infinitely many elements of M that are *not* in the range of f^0, there will be a one-to-one correspondence g between M^1 and such elements of M. We cannot just take g as the denotation of $\ddagger^1$, because (4a) requires any element of M^1 that was already present in M^0 to be assigned the same object by the functions serving as denotations of $\ddagger^0$ and $\ddagger^1$. So instead, take as the denotation of $\ddagger^1$ the function f^1 that sends an element A of M^1 to $f^0(A)$ if A is in M^0, and to $g(A)$ if not. This gives a model of $T_4 = \wp^2$V.

Note that there automatically will be infinitely many elements of M that are in the range neither of f^0 nor of f^1, namely, the values $g(A)$ for the infinitely many A in M^1 that were already in M^0. The process can be iterated indefinitely, and applies to polyadic as well as monadic second-order logic, and is adaptable to theories with Hume's principle in place of Law V, so we have the following result:

> ***Restricted Heck Consistency Theorem.*** Ramified predicative second-order logic with Law V (or Hume's principle) as an axiom is consistent.

The qualifier "restricted" appears in the name of the theorem because the systems for which Heck proved consistency are *stronger* than those in the PV series, and hence his consistency result is stronger than what has just been stated. More will be said on this topic towards the end of this chapter.

2.5 Consistency: Proof-Theoretic Proofs

Let L_0 be the language with only object variables x, y, z, … and with no non-logical relation or function symbols, but just the identity symbol =. Let T_0 be the theory whose axioms are just the formulas $\exists_k x(x = x)$ asserting that there exist at least k objects, for all $k \geq 1$. How do we know that T_0 is consistent, that no contradiction can be deduced from it?

The model theorist will say, "Any infinite set, for instance the set {0, 1, 2, …} of all natural numbers, will be a model of

T_0," leaving tacit "and so by the soundness theorem, T_0 is consistent." The proof theorist might say, "Because any deduction of a contradiction from T_0 would use only finitely many of its axioms. But for each n, we can explicitly exhibit a finite model $\{1, \ldots, n\}$ of all the $\exists_k x(x = x)$ for $k \leq n$." Here we have in miniature the difference between the "infinitistic" methods of model theory, which resist formalization in any of the arithmetical theories discussed in our survey in the preceding chapter, and the "finitistic" methods of proof theory, which generally do not require for their formalization anything beyond Q_k, for $k = 4$ or not much larger (and in this instance, do not even require so much).

In this section I will draw on three metatheorems that are known to have finitistic proofs, and with their aid will first give finitistic proofs of a number of further lemmas, and then give finitistic proofs of the consistency of PV, P^2V, and PHP. I have already explained how the existence of such consistency proofs implies (by considerations related to Gödel's second incompleteness theorem) that we are not going to be able to go far up the scale of theories Q_2, Q_3, Q_4, … in interpreting weak arithmetics in predicative Fregean theories.

The three metatheorems we will be needing are *Shoenfield's theorem*, which we met most recently in the discussion of zigzag or Wang-style set theories, the *Löwenheim-Behmann theorem*, which we met early in this chapter when discussing how much set theory we get from simple predicative second-order logic plus Law V, and *Craig's interpolation theorem* for first-order logic. Craig's theorem tells us that if ψ is logically deducible from ϕ, then there is a formula θ, called an *interpolant*, such that ψ is logically deducible from θ and θ is logically deducible from ϕ, and θ involves only vocabulary common to ϕ and ψ, which is to say that θ involves no non-logical symbols other than ones that occur *both* in ϕ *and* in ψ. Textbook presentations of this and the other two results—when textbooks present them at all—tend to give infinitistic model-theoretic proofs for them, because these are the easiest proofs to give. But in each case there is an historically early, though more difficult, finitistic proof available in the literature.[14]

The series of further lemmas we will be needing all pertain to first-order theories. The first in the series does not require any high-powered metatheorems. It is as follows. Let T be a theory in a language L. Call a formula $\phi(x)$ of L a *splitting formula* for T if for each $k \geq 1$ the formula $\exists_k x\phi(x)$ is a theorem of T. Let L^* add a one-place relation symbol P_0 and let T^* add the axioms

$$(1) \qquad \sim\exists x(P_0 x \;\&\; \sim\phi(x))$$

$$(2)_k \qquad \exists_k x\, P_0(x) \;\&\; \exists_k x(\sim P_0(x) \;\&\; \phi(x))$$

for each $k \geq 1$. (Note that $(2)_m$ implies $(2)_n$ for any $n < m$.) Then T^* is called the *splitting extension* of T for ϕ. The *splitting lemma* says that a splitting extension of a consistent theory is consistent.

Model-theoretically, this is trivial. Since T is consistent, it has a model for which, by the axioms $\exists_k x\phi(x)$, the set of objects satisfying $\phi(x)$ must be infinite. Any infinite set may be divided into a pair of disjoint infinite subsets, and taking one of the pair as the denotation of P_0, we get a model of the axioms (1) and (2) of T^*.

Proof-theoretically, one may proceed as follows (following Kit Fine). Since $\exists_1 x\phi(x)$, which is to say $\exists x\phi(x)$, is a theorem of T, we obtain a conservative extension if we add a new constant c_0 and the axiom $\phi(c_0)$. Since $\exists_2 x\phi(x)$ is a theorem of T, so is $\exists x(\phi(x) \;\&\; x \neq c_0)$, and we obtain a conservative extension if we add a new constant c_1 and the axioms $\phi(c_1)$ and $c_1 \neq c_0$. For any finite m, continue this process for $2m$ stages to obtain a conservative extension T' in a language L' with $2m$ new constants c_i for $i < 2m$, and the new axioms $\phi(c_i)$ for all i and $c_i \neq c_j$ for all $i \neq j$. Taking $P_0 x$ to abbreviate the disjunction of the identities $x = c_i$ for i even, we get an interpretation in T' of the theory that adds to T axiom (1) and the axioms $(2)_n$ for $n \leq m$, showing that extension to be consistent. Thus any finite subset of the axioms of T^* is consistent, and hence so is T^* itself.

In the remaining lemmas let us write $\exists_k\phi$ for $\exists_k x\phi(x)$, and similarly in other simple cases (so that, for instance, (1) and $(2)_k$ above would be written $\sim\exists(P_0 \;\&\; \sim\phi)$ and $\exists_k P_0 \;\&\; \exists_k(\sim P_0 \;\&\; \phi)$ respectively).

THE SECOND LEMMA we will be needing is as follows. Again let T be a theory in a language L. Formulas $\phi(x)$ and $\psi(x)$ are called *equinumeration formulas* for T if $\sim\exists(\phi\&\psi)$ is deducible from T and $\exists_k\phi$ and $\exists_k\psi$ are deducible from T for each k. Note that $\exists_k(\phi\&\sim\psi)$ and $\exists_k(\sim\phi\&\psi)$ follow immediately. Let L^* add a one-place function symbol f, and let T^* add as an axiom the conjunction $\chi(\phi, \psi)$ of the following:

(3) $\quad\forall x(\phi(x) \to \psi(f(x)))$

(4) $\quad\forall x_1\forall x_2(\phi(x_1) \;\&\; \phi(x_2) \to (f(x_1) = f(x_2) \leftrightarrow x_1 = x_2))$

(5) $\quad\forall y(\psi(y) \to \exists x(\phi(x) \;\&\; f(x) = y))$

So $\chi(\phi, \psi)$ says that f establishes a one-to-one correspondence between the x such that $\phi(x)$ and the x such that $\psi(x)$. This T^* is called the *equinumeration extension* of T for ϕ and ψ. The *equinumeration lemma* says that an equinumeration extension of a consistent theory is consistent.

Model-theoretically, this is again an easy consequence of the Löwenheim-Skolem theorem. Proof-theoretically, it will be rather more work. To begin with, since we obtain a conservative extension if we introduce one-place predicates, letting $P_1(x)$ and $P_2(x)$ abbreviate $\phi(x)$ and $\psi(x)$, we may without loss of generality assume that from the beginning the formulas $\phi(x)$ and $\psi(x)$ are atomic formulas $P_1(x)$ and $P_2(x)$. We now assume the lemma fails, in order to derive a contradiction. The failure of the lemma means that adding $\chi(P_1, P_2)$ to T produces an inconsistency, which is to say that $\sim\chi(P_1, P_2)$ is deducible from T. The Craig interpolation theorem then gives us a θ such that θ is deducible from T, $\sim\chi$ (P_1, P_2) is deducible from θ, and θ involves only $=$ and P_1 and P_2.

The Löwenheim-Behmann theorem then implies that θ may be taken to be a disjunction of conjunctions of statements about the numbers of objects satisfying each of the four combinations of P_1 and P_2. Reviewing the statement of that metatheorem, we can say more precisely that for some n, each disjunct says for each of the four combinations $(\sim)P_1\&(\sim)P_2$ either that it holds for exactly m objects, for some $m < n$, or that it holds for at least n objects. Using the fact that T implies $\sim\exists(P_1\&P_2)$, which is to say $\exists_0!(P_1\&P_2)$, we may drop any disjunct that says $\exists_m!(P_1\&P_2)$ for any $m > 0$, as well as any that says $\exists_n(P_1\&P_2)$, and the result will be a disjunction that is still implied by θ and that is logically stronger and hence still implies $\sim\chi(P_1, P_2)$. So we could have taken it for our interpolant instead, and thus we may assume that from the beginning all disjuncts of θ say $\exists_0!(P_1\&P_2)$ rather than $\exists_m!(P_1\&P_2)$ for any $m > 0$ or $\exists_n(P_1\&P_2)$. Similarly, using the fact that $\exists_n(P_1\&\sim P_2)$ and $\exists_n(\sim P_1\&P_2)$ are deducible from T, we may assume that all disjuncts of θ say $\exists_n(P_1\&\sim P_2)$ and $\exists_n(\sim P_1\&P_2)$. Thus θ is a disjunction whose disjuncts are all of the following form, where the blank may be filled in either by $\exists_m!(\sim P_1\&\sim P_2)$ for some $m < n$ or by $\exists_n(\sim P_1\&\sim P_2)$:

(6) $\quad \sim\exists(P_1\&P_2)\ \&\ \exists_n(P_1\&\sim P_2)\ \&\ \exists_n(\sim P_1\&P_2)\ \&\ \_\_\_\_$

Let us call the formula of form (6) with $\exists_k!(\sim P_1\&\sim P_2)$ in the blank σ_k, and the one with $\exists_k(\sim P_1\&\sim P_2)$ in the blank τ_k. Thus θ is a disjunction with disjuncts from among the formulas σ_m for $m < n$ and the formula τ_n.

Since the disjunction θ implies $\sim\chi(P_1, P_2)$, so does each disjunct, and as there is at least one disjunct, it follows that $\sim\chi(P_1, P_2)$ is implied by some σ_m for $m < n$ or by τ_n. Since $\exists_n(\sim P_1\&\sim P_2)$ is implied by $\exists_n!(\sim P_1\&\sim P_2)$, τ_n is implied by

σ_n, and so we may say more simply that $\sim\chi(P_1, P_2)$ is implied by σ_m for some $m \leq n$. But this is absurd, since it is possible to exhibit explicitly a finite model of σ_m & $\chi(P_1, P_2)$ with $2n + m$ objects, none satisfying $P_1\&P_2$, n satisfying $P_1\&\sim P_2$, n satisfying $\sim P_1\&P_2$, and m satisfying $\sim P_1\&\sim P_2$, and with a function for f that switches the first n objects with the second n objects and leaves the last m objects alone.

The splitting and equinumeration lemmas imply a number of further lemmas. Again let T be a theory in a language L. Let $\phi(x)$ be a splitting formula for T. Let L^* add a one-place relation symbol f and let T^* add as an axiom the conjunction of the following:

(7) $\forall x\phi(f(x))$

(8) $\forall x\forall y(f(x) = f(y) \leftrightarrow x = y)$

(9) $\exists_k y(\phi(y) \,\&\, \sim\exists x(f(x) = y))$ for all $k \geq 1$

So T^* says *all* objects can be mapped one-to-one into the objects for which ϕ holds, with infinitely many such objects left over. This T^* is called the *injection extension* of T for ϕ. The *injection lemma* says that an injection extension of a consistent theory is consistent.

To see this, we may assume ϕ is a predicate P. Apply splitting to P to get P_0, and splitting to P_0 to get P_{00}. Apply equinumeration twice to get a one-to-one correspondence g between $\sim P_{00}$ and P_{00} and a one-to-one correspondence h between P_{00} and $P_0\&\sim P_{00}$. Define $f(x)$ to be the unique y such that either $\sim P_{00}x$ and $y = g(x)$ or $P_{00}x$ and $y = h(x)$, and we are done.

Recalling that many-sorted theories are reducible to one-sorted theories by introducing appropriate one-place relation symbols, all three lemmas mentioned so far have many-sorted versions.

Again let T be a consistent theory in a language L. Let $\phi(x)$ be a splitting formula for T and let $\psi(x)$ and $\theta\,(x, y)$ be formulas such that it is deducible from T that $\exists x\psi(x)$ and that θ

defines an equivalence relative to ψ. In other words, T has the following as theorems:

(10) $\forall x(\psi(x) \rightarrow \theta(x, x))$

(11) $\forall x \forall y(\psi(x) \,\&\, \psi(y) \rightarrow (\theta(x, y) \rightarrow \theta(y, x)))$

(12) $\forall x \forall y \forall z(\psi(x) \,\&\, \psi(y) \,\&\, \psi(z) \rightarrow (\theta(x, y) \,\&\, \theta(y, z) \rightarrow \theta(x, z)))$

Let L^* be as in the injection lemma, and let T^* add as axioms the following:

(13) $\forall x(\psi(x) \rightarrow \phi(f(x)))$

(14) $\forall x \forall y(\psi(x) \,\&\, \psi(y) \rightarrow (f(x) = f(y) \leftrightarrow \theta(x, y)))$

and (9) as above for all $k \geq 1$. So T^* says that f maps the objects for which ψ holds to those for which ϕ holds, sending equivalent and only equivalent arguments to the same values, with infinitely many objects for which ϕ holds left over. This T^* is called the *representative extension* of T for ϕ and ψ and θ. The *representatives lemma* says that a representative extension of a consistent theory is consistent.

We may suppose L has at least one constant a for which T proves $\psi(a)$, for if not, one can be consistently (in fact, conservatively) added, since T proves $\exists x\psi(x)$. We may then introduce an auxiliary language $L^\dagger$ which adds an extra sort $\xi, \upsilon, \zeta, \ldots$ of variables, and a new one-place function symbol $f^\dagger$ taking arguments of sort x and giving values of sort ξ, and an auxiliary theory $T^\dagger$, which adds the following as axioms:

(15) $\forall x(\sim\psi(x) \rightarrow f^\dagger(x) = f^\dagger(a))$

(16) $\forall x \forall y(\psi(x) \,\&\, \psi(y) \rightarrow (f^\dagger(x) = f^\dagger(y) \leftrightarrow \theta(x, y)))$

It is fairly easily seen that $T^\dagger$ is consistent. Model-theoretically, as the domain of the new sort of variables we may take equivalence classes of objects for which ψ holds under the equivalence given by θ, with $f^\dagger$ denoting the function taking each object for which ψ holds to its set of equivalents, and to all others the set of equivalents of the object denoted by a. Proof-theoretically, there is a direct interpretation of $T^\dagger$ in T, associating an x variable to each ξ variable, interpreting $\forall\xi(\ldots)$ as $\forall x(\psi(x) \rightarrow \ldots)$, interpreting $f^\dagger(x)$ as the unique y such that either $\sim\psi(x)$ and $y = a$ or $\psi(x)$ and $y = x$, and interpreting $\xi = \upsilon$ as $\theta(x, y)$, where x and y are the x-variables associated to the ξ-variables ξ and υ.

Let $L^{\dagger\dagger}$ add a one-place function symbol $f^{\dagger\dagger}$ taking arguments of sort ξ and giving values of sort x, and let $T^{\dagger\dagger}$ add the following axioms:

(17) $\forall\xi\phi(f^{\dagger\dagger}(\xi))$

(18) $\forall\,\xi\forall\upsilon(f^{\dagger\dagger}(\xi) = f^{\dagger\dagger}(\upsilon) \leftrightarrow \xi = u)$

(19) $\exists_k y(\phi(y)\ \&\ \sim\exists\,\xi\,(f^{\dagger\dagger}(\xi) = y))$ for all $k \geq 1$

Then $T^{\dagger\dagger}$ is consistent by an appropriate many-sorted version of the injection lemma. Then T^* can be interpreted in $T^{\dagger\dagger}$ simply by defining $f(x)$ as $f^{\dagger\dagger}\,(f^\dagger\,(x))$.

A n-fold *equivalence* is a $2n$-place relation E with the properties analogous to those of an equivalence. For example, a triple equivalence would be a 6-place relation with the following properties:

(20) $Rx_1x_2x_3x_1x_2x_3$

(21) $Rx_1x_2x_3y_1y_2y_3 \rightarrow Ry_1y_2y_3x_1x_2x_3$

(22) $Rx_1x_2x_3y_1y_2y_3\ \&\ Ry_1y_2y_3z_1z_2z_3 \rightarrow Rx_1x_2x_3z_1z_2z_3$

Here (20), (21), (22) are just reflexivity, symmetry, and transitivity, in triplicate. The representatives lemma holds also for n-fold equivalences. Model-theoretically, the domain of the auxiliary variables ξ would be the set of sets of equivalents of n-tuples under the equivalence on n-tuples of objects associated in the obvious way with the n-fold equivalence on objects. Proof-theoretically, the interpretation would be one in which a single quantifier $\forall\xi$ is replaced by a triple of quantifiers $\forall x_1 \forall x_2 \forall x_3$ and then appropriately relativized. (This is a more general notion of interpretation than we have considered so far.) Further bells and whistles could be added, but we have now enough machinery for a number of consistency proofs.

A FINITISTIC, proof-theoretic proof of consistency for P^2V can now be obtained from the infinitistic, model-theoretic proof of consistency for $\wp^2V$ in the preceding section, by substituting appeal to Shoenfield's theorem for appeal to the existence of the power set, and appeal to Craig's theorem and the Löwenheim-Behmann theorem via the lemmas above for appeal to the Löwenheim-Skolem theorem.

To begin, let L_0 and T_0 be as at the beginning of this section. So L_0 has no non-logical symbols, and T_0 has as axioms $\exists_k x(x = x)$ for all k. Let L_1 add first round concept variables $X^0, Y^0, Z^0, \ldots$ to L_0 and with them atomic formulas of the type X^0x, and let T_1 add predicative comprehension axioms to T_0. Then T_1 is consistent by Shoenfield's theorem.[15] Let L_2 add the symbol $\ddagger^0$ and let T_2 add the following axiom:

(23) $\ddagger^0 X^0 = \ddagger^0 Y^0 \leftrightarrow X^0 \equiv Y^0$

(24) $\exists_k x \sim \exists X^0(x = \ddagger^0 X^0)$ for all $k \geq 1$

Then T^2 is consistent by an appropriate two-sorted version of the injection lemma (the roles of ψ and ϕ being taken simply by the sortal predicate for sort x and its negation, respectively). We have seen that instances of comprehension for formulas without bound X^0-variables but with the symbol $\ddagger^0$ follow from the instances without bound X^0-variables or the symbol $\ddagger^0$. So we have PV, plus the extra axioms (24).

Now we repeat the process, with a slight modification. Let L_3 add second-round concept variables $X^1, Y^1, Z^1, \ldots$ and let T_3 add predicative comprehension. Then T_3 is consistent by

Shoenfield's theorem. Let L_4 add a new function symbol † and let T_4 add the following axioms:

$$\dagger X^1 = \dagger Y^1 \leftrightarrow X^1 \equiv Y^1 \tag{25}$$

$$\dagger X^1 \neq \ddagger^0 X^0 \tag{26}$$

Then T^4 is consistent by an appropriate version of the injection lemma (with the role of ψ being taken this time by $\sim\exists X^0(x = \ddagger^0 X^0)$, using the axioms (24) above). Now † is only an auxiliary, and not yet quite what we want. But it will be provable in T_4 that for every X^1 there exists a unique x for which the following disjunction holds:

$$\begin{aligned}&\exists X^0(\forall z(X^1 z \leftrightarrow X^0 z) \;\&\; x = \ddagger^0 X^0) \vee \\ &(\sim\exists X^0 \forall z(X^1 z \leftrightarrow X^0 z) \;\&\; x = \dagger X^1)\end{aligned} \tag{27}$$

This x may be called $\ddagger^1 X^1$, and then we will have all the axioms of $\mathrm{P}^2\mathrm{V}$.

The proof for PHP, $\mathrm{P}^2\mathrm{HP}$ is essentially the same, with an appeal to the representatives lemma (of which the injection lemma is the special case where the equivalence involved is identity) in place of the injection lemma. Both with Law V as axiom and with HP as axiom, the process can be iterated indefinitely, through the single, double, triple predicative theories and on to their union, the full ramified predicative theory.

As mentioned earlier, the proof of Gentzen's cut-elimination theorem, from which all our other lemmas were derived in a fairly elementary way, requires the superexponential function, and is known to be formalizable in $Q_4 = I\Delta_0(\text{superexp})$ = superexponential functional arithmetic. It follows—though I have hardly dotted every *i* or crossed every *t* in the proof—that the consistency of $\mathrm{P}^n\mathrm{V}$ is provable for each $n = 1, 2, 3, \ldots$ in Q_4. The proof of the consistency of their union $\mathrm{P}^\omega\mathrm{V}$, which is to say, the proof that for every n the theory $\mathrm{P}^n\mathrm{V}$ is consistent, requires the assumption that for every n the superexponential function can be iterated n times, which is to say, requires the superduperexponential or super^{2}exponential function and $Q_5 = I\Delta_0(\text{super}^2\text{exp})$ = super^{2}exponential functional arithmetic. Thus the results of this section may be summed up as follows:

> ***Finitist Refinement of the Restricted Heck Consistency Theorem***. The consistency of simple, double, triple, and so on, monadic predicative second-order logic with Law V (or Hume's Principle) as an axiom is provable in superexponential functional arithmetic. The consistency of ramified monadic predicative second-order logic with Law V (or Hume's Principle) as axiom is provable in super^{2}exponential functional arithmetic.

2.6 Variable-Binding Term-Forming Operators

It remains to comment on some predicative Fregean theories where the technical issues have not yet been wholly resolved. It is only a slight exaggeration to say that all the work reported thus far in this chapter was inspired by a single paper of Heck (1996), which both proved the consistency of a ramified predicative Fregean theory, and proved the interpretability of Q in the simple predicative part thereof. Heck's systems are more complicated than any considered up to this point. Owing to their extra strength, consistency is harder to prove for them, and so far no one has produced fully worked-out finitistic consistency proofs for them.[16]

My first task must be to describe the key feature that distinguishes Heck's and related systems from all those considered so far. Generally the systems in question can*not* be construed as many-sorted first-order theories, since they allow directly, in their primitive notation, an operator {:} that can be applied to a formula $\phi(x)$ to bind the object variable x and form a term $\{x: \phi(x)\}$ that can be substituted for object variables. The operator may itself occur in the formula $\phi(x)$, so we have set terms of all ranks.

In this respect Heck's and related theories are closer to the letter of Frege than the theories considered so far, since Frege used what amount to variable-binding term-forming operators. In another respect they are arguably farther from the spirit of Frege, according to whose subordination principle concepts are primary, and the sets that are their extensions, derivative. For these systems allow a set term $\{x: \phi(x)\}$ even when they do not allow the corresponding concept, which is to say, even when they do not allow the corresponding instance of comprehension $\exists X \forall x(Xx \leftrightarrow \phi(x))$.

Interpretability in a weaker system is a stronger result than interpretability in a stronger system, and in this sense our results on interpretability of Q in predicative systems are sharpenings of Heck's result on the interpretability of Q. Heck's results remain of interest, apart from their role in inspiring later work, mainly because Heck was not simply looking to see whether some system of arithmetic could in one way or another be interpreted in his predicative theory, but rather was looking to see if it could be interpreted *using Frege's definitions of zero, successor, and natural number*. Not to put too fine a point on it, Heck found that we may follow Frege's actual chain of deductions, except in his proof of the existence of successors. There recourse has to be had to some non-Fregean way of proceeding if we are to get Q.

Consistency for a stronger system is a stronger result than consistency for a weaker system, and our results in the preceding section are of interest only because they were proved proof-theoretically and finitistically, while Heck's consistency proof is model-theoretic and infinitistic. So much for the relationship between Heck's results and those expounded so far in this chapter.

Now to describe the system Heck proves consistent. Begin with L_0, the first-order language with identity = but no non-logical symbols, and T_0, the theory with no non-logical axioms. Let $L_0{}^*$ add a variable-binding, term-forming operator {:} which attaches to any formula $\phi(x, \boldsymbol{u})$ with at least one free variable x and possibly a whole string of other free variables $\boldsymbol{u}$, to form a term $\{x: \phi(x, \boldsymbol{u})\}$ that can be substituted for object variables. (In formulas containing this term, x counts as bound, any and all parameters $\boldsymbol{u}$ as still free.) Let $T_0{}^*$ add as axioms the following:

$$(1) \qquad \forall \boldsymbol{u} \forall \boldsymbol{v}(\{x: \phi(x, \boldsymbol{u})\} = \{x: \psi(x, \boldsymbol{v})\} \leftrightarrow \forall x(\phi(x, \boldsymbol{u}) \leftrightarrow \psi(x, \boldsymbol{v})))$$

or suppressing parameters as usual, the following:

$$(1') \qquad \{x: \phi(x)\} = \{x: \psi(x)\} \leftrightarrow \forall x(\phi(x) \leftrightarrow \psi(x))$$

This $T_0{}^*$ is essentially what Parsons (1987) is pleased to call the "first-order portion of the *Grundgesetze*," though it is not literally a first-order theory, owing to the presence of the operator {:}. It is for this theory that Parsons supplied a model-theoretic consistency proof.

A finitistic proof of the consistency of $T_0{}^*$ is implicit in our earlier results, since we have in effect seen that the part of $T_0{}^*$ involving only set terms of rank n is interpretable in $\mathrm{P}^{n+1}\mathrm{V}$, and that the consistency of all $\mathrm{P}^n\mathrm{V}$ is provable finitistically. Since any deduction of a contradiction in $T_0{}^*$ would involve only finitely many formulas, and therefore only terms up to some finite rank n, the consistency of $T_0{}^*$ follows.

Before going any further, it may be remarked that the set theory UST with extensionality, null set, and pairing can be interpreted in $T_0{}^*$. Relativize quantifiers to the following formula:

$$(2) \qquad y = \{x: x = x\} \vee \exists u \exists v(y = \{x: x = u \vee x = v\})$$

Interpret $z \in y$ by the following formula:

$$(3) \qquad \exists u \exists v(y = \{x: x = u \vee x = v\} \,\&\, (z = u \vee z = v))$$

The interpretability of UST means that as soon as we add first-round concept variables and predicative comprehension for them, even without allowing any more set terms, Q and hence Q_2 will be interpretable by the results of this chapter (and similarly for Q_3 if we go on to a second round).[17]

Now let L_1 add first-round concept variables X^0, Y^0, Z^0, … and allow the formation of set terms $\{x: \phi(x)\}$ for ϕ that may contain free first-round concept variables but do not contain bound first-round concept variables. Let T_1 add predicative comprehension:

$$(4) \qquad \exists X^0 \forall x(X^0 x \leftrightarrow \phi(x))$$

provided $\phi(x)$ *contains no bound concept variables*

along with the new instances of (1) for the new set terms allowed by L_1. Let L_1* allow set terms $\{x: \phi(x)\}$ also for ϕ that may contain bound first-round concept variables, and let T_1* add the new instances of (1) for the new set terms allowed by L_1*.

This T_1* is what Heck takes as his "simple predicative" theory. Obviously, introducing second-round concept variables X^1, Y^1, Z^1, …, we can go on to form L_2 and T_2, L_2* and T_2*, and then continue the series. Heck gives a model theoretic consistency proof for T_1* that generalizes to the other theories in the series.

There are two ingredients to the proof. The first ingredient is an improved version of the Parsons proof of the consistency of T_0*, which not only is used to get the consistency of T_0* but also is used to get the consistency of T_n* once the consistency of T_n is established. The second ingredient, which gives the consistency of T_{n+1} once the consistency of T_n^* is established, is an extended version of Shoenfield's theorem.

This result may be stated as follows. Suppose we have a language L that is first order *except for allowing a variable-binding term-forming operator* {:}, and have a theory T that has an axiom scheme

(5) $$—\phi(x)—\psi(x)—$$

where $\phi(x)$ and $\psi(x)$ may be any formulas of L, perhaps containing free variables besides x, and *may occur in* (5) *through terms* $\{x: \phi(x)\}$ *and* $\{x: \psi(x)\}$ *as well as directly*. Let L' add concept variables X, Y, Z, and let T' add predicative comprehension

(6) $$\exists X \forall x(Xx \leftrightarrow \phi(x))$$
provided ϕ *contains no bound concept variables*

and let T' add also all instances of (5) for ϕ and ψ containing no bound concept variables. Then the extended Shoenfield theorem says that T' is consistent. Shoenfield's original theorem is the same *without the italicized parts*. Heck shows that the usual model-theoretic proof of Shoenfield's original theorem can be elaborated to handle variable-binding term-forming operators.[18]

A SOMEWHAT STRONGER consistency result than ours on the consistency of PV was obtained by Wehmeier (1999). His system is essentially our PV—among early writers on these matters Wehmeier was the only holdout against the tradition of allowing a variable-binding, term-forming operator {:}—but with the following strengthened form of the comprehension axiom, called Δ_1^1 -comprehension:

(7) $$\forall x(\exists U\phi(x, U) \leftrightarrow \forall U\psi(x, U)) \rightarrow \exists X\forall x(Xx \leftrightarrow \exists U\phi(x, U))$$
provided ϕ *and* ψ *contain no bound concept variables*

This says that if a formula $\exists U\phi$ having a single, initial existential quantifier and a formula $\forall U\psi$ having a single, initial universal concept-quantifier—Σ^1_1; and Π^1_1 formulas—hold of the same objects, then they determine a concept. If ϕ does not contain the variable U, the antecedent of (7) is understood to hold vacuously for $\phi = \psi$. Thus Δ^1_1-comprehension subsumes predicative comprehension as a special case.[19]

An analogous strengthening of Heck's consistency result for $T_1{}^*$ was subsequently obtained by Ferreira and Wehmeier (2002). Their system is essentially Heck's $T_1{}^*$ with Δ^1_1-comprehension.[20] In connection with Δ^1_1-comprehension it should be noted that the set of all sets x that are not elements of themselves has a Σ^1_1 definition, which we have seen in several variant notations before:

(8) $$\exists X(x = \{y: Xy\} \mathbin{\&} \sim Xy)$$

Similarly, the set of all objects that are not elements of themselves has a Π^1_1 definition as follows:

(9) $$\forall X(x = \{y: Xy\} \rightarrow \sim Xy)$$

So either Σ^1_1-comprehension or Π^1_1-comprehension would lead to inconsistency. And so would Δ^1_1-comprehension if (8) and (9) were equivalent, and in particular, if every object were a set.

To put the matter another way, Δ^1_1-comprehension for first-round concepts implies that *not* every object is the extension of a first-round concept. In fact, Wehmeier, who noted this point, went on to show by an extension of the argument that for every m, Δ^1_1-comprehension implies the existence of at least m objects that are not such extensions. He describes the theory as implying the existence of an infinity of "non-logical" objects.

If it were the case, as this description suggests, that Δ^1_1-comprehension is something one can't consistently assume without assuming for each *m* that there exist at least *m* "non-logical" objects, then indeed the theory would involve an assumption uncomfortably close to Russell's axiom of infinity, and completely contrary to the Fregean idea that *logical deduction*, not *arbitrary postulation*, should give us an infinity of objects.

Justified though Wehmeier's conclusion may be when the simple Δ^1_1-comprehension theory is viewed in isolation, from the ramified perspective, where the theory is only the first of what is to be an endless series involving more and more rounds of concepts, it seems premature (like the conclusion one might draw from a simple predicative theory considered in isolation that certain formulas are being allowed into the language that do not determine concepts). The objects stigmatized as "non-logical," though indeed not the extensions of first-round concepts, may nonetheless be "logical," in that nothing excludes their being extensions of second-round concepts. (We know there will be extensions of second-round concepts that are not the extensions of any first-round concepts.) Indeed, in a fully ramified theory with infinitely many rounds of concepts, it is not even possible to *state* that an object is "non-logical" in the sense of not being the extension of a concept of any round, much less prove that there are any "non-logical" objects, since there is no quantification over concepts of all rounds.

Be all that as it may, the Ferreira-Wehmeier proof uses rather more sophisticated model-theoretic techniques than the Heck proof, and is correspondingly further away from being finitistic.[21] We do not yet have a definite upper bound on how large a fragment of arithmetic the Ferreira-Wehmeier system can interpret. Thus it certainly has not yet been rigorously proved that one cannot get Q_4 in a predicative Fregean system, for instance. Despite such loose ends, however, I believe no one working in the area seriously expects to be able to get very much further in the sequence Q_m while working in predicative Fregean theories of whatever kind. It is time, therefore, to have a look at the impredicative option.

3

Impredicative Theories

The preceding chapter was devoted to systems that assume the existence of extensions for whatever concepts are allowed to exist, but that do not allow formulas involving quantification over absolutely all concepts to determine concepts, or that simply do not allow formulas involving quantification over absolutely all concepts. The present chapter is devoted to theories that take the opposite tack, and accept full second-order logic, with quantification over all concepts freely admitted in formulas, and with all formulas assumed to determine concepts, but that restrict the assumption of the existence of extensions, or replace it by the assumption of the existence of abstracts for some equivalence other than coextensiveness.

The work to be noted includes contributions by Crispin Wright (1983), Neil Tennant (1987), J. L. Bell (1999), Bob Hale (2000), Kit Fine (2002), Øystein Linnebo (2004), and again as in the preceding chapter, Heck and Boolos.

3.1 Frege Arithmetic and Less

Theories of the kind to be considered in this chapter generally permit the interpretation of much larger fragments of classical mathematics than did the predicative theories considered in the preceding chapter, which is why I have deferred consideration of them until this one. But chronologically, impredicative abstractionist theories actually appeared in the literature first, and it may be well to say a bit about the history. Neo-Fregeanism as it has developed over the past two decades was launched by the publication of Crispin Wright's book, *Frege's Conception of Numbers as Objects*, in 1983. That book for the first time brought together two observations: first, that the Russell paradox does not arise if one drops Law V and works only from Hume's principle; and second, that working from Hume's principle one can derive arithmetic.

Each of these observations had been made separately, at least in passing, a decade or two earlier. On the second point, Charles Parsons (1965, §VI), after recalling Frege's definition of the number of a concept, which Parsons writes $N_x . Fx$, after recalling Frege's derivation from this definition of Hume's principle, which Parsons calls (A), and after recalling Frege's definitions of zero, successor, and natural number in terms of the notion of number, makes the following remark:

> From these, the Peano axioms can be proved; it is not necessary to use any axioms of set existence except in introducing terms of the form '$N_x . Fx$' and in proving (A), so that the argument could be carried out taking (A) as an axiom.

Wright on this point went further than Parsons, in actually sketching a derivation of the Peano postulates, though the sketch stops short of being a fully rigorous deduction, such a deduction being promised for a later joint publication with Neil Tennant.

On the first point, Peter Geach (1975, 446–47) says the following by way of criticism of a remark of Dummett's:

> But we have on the contrary every reason to believe that there is a totality closed under the operation of mapping arbitrary predicates defined over it onto *cardinal numbers*. Let us consider the

> domain consisting of the natural numbers together with $\aleph_0$. To any predicate defined over this domain there attaches either a natural number or $\aleph_0$, as the number of objects of which the predicate is true; e.g., the number 4 attaches to the predicate '— is prime and less than 10', the number $\aleph_0$ to the predicate '— is even'. We have no need to go outside the domain to find cardinal numbers attaching to these predicates.

Wright on this point did not go as far as Geach, for he did not offer a model to show HP consistent, but merely indicated how the attempt to derive the Russell paradox breaks down, and conjectured that the system with HP rather than Law V is consistent.

What Wright did that had not been done before was to bring together the two observations and offer, at least as a conjecture with some formal support, the observation that *arithmetic can be derived in a* consistent *theory consisting of second-order logic with* HP *as an additional axiom*—that is to say, in the impredicative abstractionist theory that has come to be called *Frege arithmetic* or FA. Besides this, Wright offered vigorous argumentation for the claim that such a derivation would be of substantial philosophical significance. An intensive investigation of the technical aspects of Frege arithmetic, and an intense debate over its philosophical merits and demerits, promptly followed, as well as explorations of how to move beyond Frege arithmetic to an abstractionist version of or foundation for higher mathematics.

In this chapter we will have a brief look first at abstractionism as developed by Wright, his collaborator Bob Hale, and their school, then turn to the work of Kit Fine, and to abstractionist set theory in the work of Boolos. Then I will suggest how one could go much further, at the cost of being rather less Fregean.

To RETURN TO Wright, his book left some loose ends. For one thing, the joint paper of Wright and Tennant never materialized. For another thing, the consistency of FA was stated by Wright merely as a conjecture, and though restated as fact the following year by Harold Hodes (1984), it was still left without proof. These loose ends were, however, fairly quickly tied up. On the one hand, a fully rigorous deduction of the Peano postulates in Frege arithmetic was published by Tennant on his own, though unfortunately in a locus, Tennant (1987), where it escaped the attention of most philosophers and logicians who might have been interested. On the other hand, two reviews, Hazen (1985) and Burgess (1984), pointed out that consistency follows from the existence of a simple model for FA; the model was, indeed, Geach's, but his remark had been entirely overlooked by all concerned (and remained so for almost twenty years until Saul Kripke came across it).

A fully satisfactory treatment, however, had to await the work of another early reader of Wright who had also independently rediscovered the Geach model, George Boolos. He introduced the terminology still used in discussing these matters: "Hume's principle" or "HP" (which Wright had called "$N^=$"), "Frege's theorem," "Frege arithmetic." And his beautiful paper, Boolos (1987), made a crucial new point. For he showed not only that the Peano postulates can be derived in FA (essentially by following Frege's own chain of deductions), so that second-order Peano arithmetic P^2 becomes interpretable in FA (with Frege's definitions of zero, successor, and natural number), but also that the following converse holds:

The Geach-Boolos Converse to Frege's Theorem. Frege arithmetic is interpretable in second-order Peano arithmetic.

The idea is to mimic the Geach model while working within second-order Peano arithmetic, with 0 playing the role of $\aleph_0$ and y' playing the role of y. Writing U for the *universal concept* «x: $x = x$» and $\langle y \rangle$ for the *strict segment* «x: $x < y$», we must first prove that every X is equinumerous either with U or with some $\langle y \rangle$, where these alternatives are exclusive, and y unique. We can then define the operation # by letting # X be 0 if X is equinumerous with U, and letting # X be y' if X is equinumerous with $\Sigma\langle y \rangle$. A bit more will be said about the proof below.

A series of papers by Boolos and Heck (independently and jointly) searchingly examined several possible derivations of the Peano postulates and other results from HP, both from a logical point of view and from an historical point of view, in comparison with Frege's way of proceeding in the informal sketch in the *Grundlagen* and the formal treatment in the *Grundgesetze*.[1] Here historical issues will be entirely left aside, and as to logical issues, only two will be addressed in any depth. First, can the assumptions of FA be significantly weakened without losing the Peano postulates? Second, can the assumptions of FA be strengthened in a manner that is not *ad hoc*, so as to give more of classical mathematics? In the remainder of this section I will take up the first of these two questions, turning to the second in the next section.

HUME'S PRINCIPLE provides abstracts with respect to the equivalence of equinumerosity. Such abstracts include the natural numbers, which were Frege's primary though by no means exclusive interest, as well as Cantor's transfinite cardinals, though within FA one can only *prove* the existence of the smallest of these, $\aleph_0$. It would be surprising if the existence of even one transfinite cardinal were indispensable for the derivation of the Peano postulates, and in fact it is not: the Peano postulates are still forthcoming if one restricts HP so that it provides cardinals only for finite sets.

This fact has been noted (in one form or another) independently several times.[2] Let us say that a definition of finitude is *minimally adequate* if it permits the deduction of the following two conditions (nullity and adjunction):

(1) $$\sim\exists\, xXx \rightarrow X \text{ finite}$$

(2) $$X \text{ finite} \;\&\; \forall x(Yx \leftrightarrow Xx \vee x = y) \rightarrow Y \text{ finite}$$

(Both are provable for Dedekind and Russell finitude.) Then we have the following abstract version of the result in question:

> ***The Tennant-Heck-Bell Refinement of Frege's Theorem.*** Second-order Peano arithmetic is interpretable in dyadic second-order logic with *finite* HP as an axiom, for any minimally adequate notion of finitude.

Turning from weakenings of Hume's principle to weakenings of the underlying logic, a point emphasized by Tennant (and independently rediscovered by Bell) is that the derivation of the Peano postulates does not require the law of the excluded middle $A \vee {\sim}A$ or other principles of classical logic to which intuitionists object. But intuitionist views are beyond the scope of the present study.

More relevant, obviously, is the question of how far the derivation of the Peano postulates depends on having second-order logic with impredicative comprehension. This issue was touched on in the preceding chapter, but there is more to be said about the question, which has especially been investigated by Øystein Linnebo (2004). We have already seen in the preceding chapter that with only predicative comprehension, it is possible to interpret Q and hence Q_2, but that consistency can be proved finitistically, so it is impossible to interpret Q_m for large m. Linnebo provides information about the relationship between fragments of FA and fragments of P^2 given by versions of comprehension that are impredicative but still stop well short of full comprehension.

Among these the most important are systems where comprehension is assumed for and only for Π^1_1-formulas, or formulas consisting of a string of universal second-order quantifiers followed by a formula without further second-order quantifiers. These may equally be described as systems where comprehension is assumed for and only for Σ^1_1-formulas, consisting of a string of existential second-order quantifiers followed by a formula without further second-order quantifiers. The two formulations are equivalent because once one has a concept, predicative comprehension is enough to give its complement, and Π^1_1-formulas and Σ^1_1-formulas are equivalent to the negations of each other.

Actually, some questions of formulation arise as to just what should be considered the Π^1_1-version of second-order Peano arithmetic and what should be considered the Π^1_1-version of Frege arithmetic. On the Peano side, we have seen that with full comprehension the single-axiom version (I_0) of induction is equivalent to the axiom-scheme version (I), but that when comprehension is restricted, (I) is stronger than (I_0). And in particular there are two Π^1_1-versions of second-order Peano arithmetic, the stronger Π^1_1-CA with (I), and the weaker Π^1_1-CA_0 with (I_0). The experience of mathematical logicians has shown that the latter system is the more significant foundationally. (It is the last of the Friedman systems, F_4.)

For purposes of comparison with Frege arithmetic, however, the most appropriate system is a slightly different one I will call Π^1_1-P^2 for Π^1_1-*Peano arithmetic*. This system consists of dyadic second-order logic with Π^1_1-comprehension, and axioms (Q1) and (Q2) for zero and successor. Π^1_1-P^2 is interpretable in Π^1_1-CA_0, which has only monadic second-order logic, using the coding of pairs of natural numbers by single natural numbers that is definable in terms of addition and multiplication in P^1 and hence in Π^1_1-CA_0. Since Π^1_1-CA_0 is weaker than Π^1_1-CA, so is Π^1_1-P^2. But Π^1_1-CA_0 and Π^1_1-P^2 are of equivalent strength, because we have the following:

Dedekind's Theorem, Version III. Π^1_1-CA_0 is interpretable in Π^1_1-P^2.

The interpretation called for by the theorem proceeds by defining the notion of *inductive* in the usual way (X is *inductive* if and only if we have $X0$ and have Xx' whenever Xx), and then defining the notion of *natural number* in the usual way given the notion of inductive (x is a *natural number* if we have Xx for every inductive X). This definition is expressible by a Π^1_1-formula $\nu(x)$ (with "*every* inductive X" giving the initial universal second-order quantifier). Hence by Π^1_1-comprehension there exists an N such that for all x, we have Nx if and only if x is a natural number.

And now for a subtle but important point. It follows that Π^1_1-comprehension holds *for formulas relativized to natural numbers*. For given a Π^1_1-formula $\phi(x)$, say of form $\forall X \exists y \forall z \ldots$, its relativization ϕ^ν, looks like this:

(3) $$\exists Y \forall x(\nu(x) \rightarrow (Yx \leftrightarrow \forall X \exists y(\nu(y) \;\&\; \forall z(\nu(z) \rightarrow \ldots x \ldots X \ldots y \ldots z \ldots))))$$

This is certainly not itself an instance of Π^1_1-comprehension because of the second-order quantifiers in $\nu(x)$. But (3) is a consequence of the following two instances of Π^1_1-comprehension:

(4) $$\forall U \exists Y \forall x(Yx \leftrightarrow Ux \;\&\; \forall X \exists y(Uy \;\&\; \forall z(Uz \rightarrow \ldots x \ldots X \ldots y \ldots z \ldots))$$

(5) $$\exists U \forall x(Ux \leftrightarrow \nu(x))$$

A slightly more informal way to put the point would be as follows. The fact that ν is a Π^1_1-formula means that Π^1_1-comprehension (specifically, the instance (5)) implies that the concept N of being a natural number exists or belongs to the domain of the second-order variables, and is therefore available for use as a parameter in other Π^1_1-formulas to which Π^1_1-comprehension may be applied, as in the following:

(6) $$\exists Y \forall x(Yx \leftrightarrow Nx \;\&\; \forall X \exists y(Ny \;\&\; \forall z(Nz \rightarrow \ldots x \ldots X \ldots y \ldots z \ldots))$$

A similar point is that we have induction in the single-axiom form (I_0), since if X is inductive *relative to natural numbers* (so that we have $X0$ and at least have Xx' whenever Xx *and* x *is a natural number*) then the intersection of X and N is inductive *tout court*, and every natural number by definition must fall under this intersection and hence under X.

Addition and multiplication can then be introduced as in the proof of the interpretability of Q in the preceding chapter, which required only predicative comprehension, and then we have recovered the more conventional formulation Π^1_1-CA_0 (the Friedman system F_4).[3]

On the Frege side, we have a choice between a system having a single-axiom version of HP, like PHP of the last chapter, but with Π^1_1-comprehension rather than predicative comprehension, and a system comparable to Heck's, which would allow terms $\|x\colon \phi(x)\|$ of all ranks and an axiom-scheme version HP applying to them, even for formulas ϕ for which comprehension is not assumed. The more tractable system, and the one most relevantly compared to Π^1_1-CA_0 and Π^1_1-P^2 is the former system, and it is *that* system that I will call Π^1_1-FA for Π^1_1- *Frege arithmetic*. We then have the following pair of results:

Linnebo's Refinement of Frege's Theorem. Π^1_1-Peano arithmetic is interpretable in Π^1_1-Frege arithmetic.

Converse to Linnebo's Refinement of Frege's Theorem (Refinement of the Geach-Boolos Converse to Frege's Theorem). Π^1_1-Frege arithmetic is interpretable in Π^1_1-Peano arithmetic.

Linnebo's work has an historical in addition to a logical aspect, in that what Linnebo actually proves is something stronger than what has just been stated, namely, that Π^1_1-Peano is interpretable in Π^1_1-Frege arithmetic *using Frege's definitions of zero, successor, and natural number*. Moreover, in contrast to the result in the preceding chapter on the interpretability of Q in PHP, Linnebo shows that the successor theorem (asserting the existence of a successor for every natural number) cannot be proved in predicative Frege arithmetic *if one uses Frege's definitions of zero, successor, and natural number*. This is so whether or not one tries to follow Frege's chain of deductions, and even if one works with the strengthening of predicative Frege arithmetic with the variable-binding, term-forming operator $\| : \|$. This negative result is proved by a model construction closely patterned on Heck's model construction, but bringing in an application of the Löwenheim-Behmann theorem.

Linnebo's proof of the interpretability of Π^1_1-Peano arithmetic in Π^1_1-Frege arithmetic consists essentially of going through Frege's proof of Frege's theorem and making certain adjustments to avoid dependence on instances of comprehension for formulas that are too complex. The proof of the converse of Linnebo's theorem involves similarly going through and adjusting Boolos's proof of the converse of Frege's theorem. I will give here only an outline of the proofs of the two theorems just enunciated, leaving the details as an exercise for the reader.

For the first theorem, I will go through and adjust, not Frege's proof of Frege's theorem, but rather the proof of the interpretability of Q in PHP from the preceding chapter. The disadvantage of this procedure is that we do not get Linnebo's result about what can be accomplished *using Frege's definitions*; the advantage is that it requires very little additional work, since we already know that nothing but predicative comprehension was used to get an interpretation of (Q1) and (Q2). So all that remains to be done to get an interpretation of Π^1_1-P^2 in the present context is to show that Π^1_1-comprehension still holds when formulas are relativized to the formula $\pi(x)$ expressing that x is a protonatural number. To see this, we need only review the definitions of the auxiliary notion $\lhd$ and of protonatural, which were as follows:

(7) $x \lhd y \leftrightarrow \exists X \exists Y(x = \#X \,\&\, X \subset Y \,\&\, y = \#Y)$

(8) x is protonatural $\leftrightarrow \exists X(\forall y(Xy \leftrightarrow y \lhd x)$
$\&\, x = \#X \,\&\, \sim Xx))$

The definition (7) shows that Σ^1_1-comprehension implies that the relation $\lhd$ exists and belongs to the domain of the dyadic second-order variables, making it available for use as a parameter in later definitions, notably (8). Again the definition (8) shows that Σ^1_1-comprehension implies that the concept P of being protonatural exists and belongs to the domain of the monadic second-order variables. And this is all we need show that Π^1_1-comprehension still holds when formulas are relativized to the formula $\pi(x)$ expressing that x is a protonatural number. (Compare the "subtle but important" step in the proof of Dedekind's Theorem, Version III, above.)

For the converse theorem, by Dedekind's Theorem, Version III, it is enough to show that Π^1_1-FA is interpretable in Π^1_1-CA_0. The idea of the original Boolos proof of the Geach-Boolos theorem that FA is interpretable in P^2 was indicated at the beginning of this section, immediately after the statement of that theorem. The claim is that this idea can be implemented in the present context, where we have only Π^1_1-comprehension.

Towards such an implementation, define X to be *bounded* if for some y all x such that Xx have $x < y$, and *unbounded* otherwise. Recall that in P^1 one has not only a coding of pairs of natural numbers by single natural numbers, but also a coding of finite sequences. A number coding sequence of length y that has no repetitions and whose entries are all and only the x such that Xx in effect codes a one-to-one correspondence between $\langle y \rangle$ and X. One can then prove *using only predicative comprehension* the *main lemma* that for any X, either X is unbounded, in which case X is equinumerous with the universal concept U, or there is a number coding a one-to-one correspondence between X and some $\langle y \rangle$, where these alternatives are exclusive, and y unique. It follows that for any X there is a unique z such that either X is unbounded and $z = 0$, or for some y there is a number coding a one-to-one correspondence between X and $\langle y \rangle$, and $z = y'$. Moreover, if we define $\#X$ to be this unique z, then Hume's principle will hold. And finally, there is the subtle but important point that since as just given *the definition of # involves no bound second-order variables*, Π^1_1-comprehension still holds when the defined symbol # is allowed to appear in formulas. The proof of the main lemma will be left as an exercise for the reader.[4]

The mutual interpretability result connecting Π^1_1-Peano arithmetic and Π^1_1-Frege arithmetic extends to Π^1_n for $n > 1$, and also to Δ^1_n for $n > 1$. The cases of Δ^1_1-comprehension and of Σ^1_1-separation (corresponding to the Friedman system F_3) remain open questions. Theories that have the axiom-scheme version of induction rather than the single-axiom version on the Peano side, and have the variable-binding, term-forming operator $\| : \|$ on the Frege side, also remain to be examined. But despite such loose ends, the general impression left by all the results discussed so far is that there is something irreducibly impredicative about Frege's procedure, which can be minimized but not eliminated. It follows that any philosophical position that wishes to found more than a weak fragment of arithmetic in anything like Frege's way will have to accept predicativity as not being always *viciously* circular.

Linnebo actually distinguishes two dimensions of impredicativity. One is the usual matter of allowing bound concept variables in the comprehension scheme $\exists X \forall x(Xx \leftrightarrow \phi(x))$, discussed above. But he also sees a species of impredicativity to be involved in the fact that $\#X$ is assumed to be an object, and among the very objects whose falling under or not falling under X or Y may matter to whether we have $X \approx Y$ or not.

This assumption, that numbers are themselves among the objects that can be numbered, was important for Frege's polemic against the notion that numbers are somehow connected with spatiotemporal intuition. It was also, of course, crucial to Frege's bootstrapping idea for proving the successor theorem. It is therefore not surprising that Linnebo is able to show that if we assume that numbers $\#X$ are entities of some third sort ξ, off to the side, as it were, from the objects x and the concepts X, then we cannot prove the existence of more than a very few of them. We still get 0 as the number of the empty concept, and we still get 1, courtesy of the convention that we always assume there is at least one object, and that is all.[5]

As we saw in the proof of the "representatives lemma" in the preceding chapter, when a first-order theory T in an n-sorted language L is extended by adding abstracts with respect to some equivalence on one or another of the n sorts, with a new $(n + 1)$st sort being added for the abstracts, the extended theory T' in the extended language L' is conservative over the original theory in the original language. Any formula of L provable in T' is provable in T. Thus no old questions get new answers. Moreover, there are in a sense not any new questions, either, since every formula of L' can be proved in T' to be equivalent to a formula of L.[6]

By contrast, in the case of adding numbers in the usual way, using HP, we may get new answers, since HP implies there are infinitely many objects, which the original theory may not (and does not if the original theory is simply pure second-order logic). We also get new questions. We can ask, to begin with, questions about the identity of particular numbers with particular objects of kinds recognized before the introduction of numbers.[7] Such is Frege's notorious question whether the number two is the same as Julius Caesar. We can also ask, for instance, if there are as many numbers as there are objects.[8]

How the second dimension of impredicativity, and the double non-conservativeness to which it gives rise, makes it harder to claim philosophical, and especially epistemological, importance for Frege's theorem is perhaps best brought out by means of a parable.

The geometers of Flatland had a fine geometry of the plane that was their world, but they noted a certain asymmetry in it. While any two points determine a unique line on which they both lie, not every two lines determine a unique point lying on both, since some pairs of lines are parallel and have no common point. One day, however, a visionary proclaimed that all pairs of lines *do* meet in a common point, but with this distinction, that while most pairs meet in an *ordinary* point, those pairs called parallel meet only at a *point at infinity*. The recognition of these points at infinity makes it possible to maintain the principle that any two lines determine a unique point as an exceptionless law.

Nor does the recognition of these points at infinity compromise the principle that any two points determine a unique line. Two ordinary points still determine a line in the ordinary way. An ordinary point and a point at infinity determine a line as well. For taking any line through the point at infinity, it was already a principle of the old geometry that there is a unique line through the given ordinary point that is parallel to the given line, and in the new geometry this means that it passes through the given point at infinity. As for two points at infinity, all such points lie on a common *line at infinity*.

The new infinitistic geometry soon proved to have the power to unify many disparate results of the old geometry, and to suggest new results as well, in a way that will be familiar to those acquainted with projective geometry in our own world (Hilbert's favorite example of the utility of "ideal elements"). Yet despite its utility, the new geometry was regarded with suspicion by some geometers, and more philosophers. After all, no one's eyesight is keen enough to enable them to see the infinitely distant line with its infinitely distant points from any place in the ordinary plane, nor are anyone's feet swift enough to enable them to go to these infinitely distant places and see them up close.

After controversy between those who accepted the vision of a new geometry and those who rejected it had gone on for some time, a logician arose, who with two lectures was able to put an end to the debate. During the first lecture, the logician revealed how the points at infinity, and even the line at infinity, could be regarded as entities of a new sort obtained as *abstracts* from entities of an old sort, points-or-lines, with respect to a certain equivalence, defined as follows:

$X \| Y \leftrightarrow$ X and Y are both lines and are parallel
$\vee$ X and Y are both points.

Introducing a new sort of variable ξ for abstracts with respect to this equivalence, and assuming, as is usual with abstracts, that to any X there corresponds a unique ξ in such a way that the same ξ attaches to two different Xs if and only if they are equivalent, then we may identify the *point at infinity* on a given line X as its abstract ξ, more commonly called its *direction*. And at the cost of a little artificiality, we may identify the *line at infinity* with the one remaining abstract ξ common to all points X.

During the second lecture, the logician revealed that whenever entities of a *new sort* are introduced as abstracts, everything said about them can be reinterpreted as a statement just about entities of the old sort. As a result, by the end of the second lecture, even those who had not the least inclination to believe in the literal truth of the visionary theory of the infinitely distant line and its infinitely distant points were equipped with a method by which they could reinterpret everything said by believers in such a way as to make it come out true. And this was enough to put an end to the controversy.

The logicians of Flatland had a theory of concepts, too, but there was a gap in their theory. For they found themselves unable to prove that for each finite concept there is another

> finite concept such that all the objects falling under the latter bar one are equinumerous with *all* the objects falling under the former. One day, however, a visionary announced that there were objects called *numbers*, and using certain assumptions about these numbers, the visionary was able, by a very ingenious argument, to fill in the gap.
>
> Great mathematical developments ensued, and great philosophical controversy. The logician who had settled the geometrical controversy was invited to lecture on the arithmetical question. At a first lecture, the logician revealed how the numbers could be obtained from concepts as *abstracts* with respect to equinumerosity. But there was no second lecture, and the controversy continues to this day.

It is the "second dimension of impredicativity" and the ensuing "double non-conservativeness" that are most directly responsible for the impossibility of giving a "second lecture" in the foregoing parable. But let me now finally put the topic of predicativity and impredicativity aside, and turn for good to systems with full comprehension.

3.2 Frege Arithmetic and More

The system P^2 does not bear the alternative name *analysis* without reason. It really is possible, through appropriate coding, to develop the bulk of mathematical analysis within this system, with natural, integral, and rational numbers being represented by objects x, and real and complex numbers, and open sets thereof and continuous functions thereon being represented by concepts X. Since P^2 can be interpreted in FA, it is equally true that the bulk of mathematical analysis can be squeezed into the Fregean system.

A more ambitious Fregean, however, will want to give an interpretation in which real and complex numbers are *objects*, making *arbitrary* sets of and functions on them available, at least as concepts, and thus making at least P^3 interpretable in the Fregean system based on second-order logic—and if third- or fourth- or fifth-order logic is allowed in the Fregean system, one would get P^4 or P^5 or P^6, though most recent workers stop at order two.

The possibility of moving beyond the theory of the natural numbers as developed in FA to obtain other sorts of numbers as abstracts with respect to other sorts of equivalences on objects or on concepts has especially been pursued by Bob Hale (2000) and other collaborators of Wright, mainly at the Arché Institute at the University of St. Andrews—the Scottish school, for short.

As a matter of fact, the integers, rational numbers, and real numbers can all be quite naturally introduced as abstracts. For the integers, one may consider the double equivalence of *equidistance* or equality of directed distance between natural numbers, defined to hold among x, y, u, v if and only if one of the following three alternatives holds: either both $x < y$ and $u < v$ and for some w we have both $x + w = y$ and $u + w = v$; or both $x = y$ and $u = v$; or else both $y < x$ and $v < u$ and for some w we have both $y + w = x$ and $v + w = u$. It follows from the basic laws of arithmetic that for natural numbers, any x, y stand in the equidistance relation either to $0, w$ or to $w, 0$ for some natural number w. Introducing abstracts for pairs of natural numbers with respect to equidistance, the negative integer $-w$ and the positive integer $+w$ may be identified with the abstracts for these two pairs. Arithmetical operations on integers can then be defined in terms of arithmetical operations on natural numbers.

Rational numbers may be obtained from pairs of integers x, y with $y \neq 0$ as abstracts for the double equivalence of *proportionality*. The construction is not dissimilar to that for integers, but with multiplication in place of addition.

For the reals, one may define x to be a *rational upper bound* on X if x is a rational number, and for every rational number y such that Xy we have $y \leq x$. Calling X and Y *coterminous* if every rational upper bound for either is a rational upper bound for the other, we obtain an equivalence on concepts, and may obtain real numbers as abstracts for this equivalence. (Actually, if we consider *arbitrary* X, we will obtain the *extended* real numbers, with $-\infty$ as the abstract for those X having Xy for no rational y, and $+\infty$ as the abstract for those X having Xy for arbitrarily large rational y.) This is but a slight variant presentation of Dedekind's treatment of the real numbers. For a more detailed account of it, see Shapiro (2000), and for a comparison with the approach of Hale (2000), see Wright (2000).

Innumerable variations are available. We could, for instance, get to the rationals by first introducing fractional and then introducing negative numbers, rather than the reverse as was done above. And Dedekind's treatment of the real numbers admits several alternatives, one due to Cantor, another due to Frege himself (whose treatment Hale, for one, wishes to follow, in preference to other approaches, including that outlined above).

Moreover, there is a very direct route to an interpretation of P^2, in which the theory of real numbers can be developed. After obtaining natural numbers from Hume's principle, let $N =$ «x: x is a natural number» and consider *numerical equivalence*, defined thus:

(1) $$X \equiv_N Y \leftrightarrow ((\exists x(Xx \mathbin{\&} \sim Nx) \vee \sim\exists x Xx) \mathbin{\&} (\exists x(Yx \mathbin{\&} \sim Nx) \vee \sim\exists x Yx)) \vee X \equiv Y$$

And assume we have abstracts for this equivalence, subject to what we may call *numerical Law* V or *N*V:

(2) $$\ddagger_N X = \ddagger_N Y \leftrightarrow X \equiv_N Y$$

Obviously, these abstracts provide surrogates for sets of natural numbers (with the one abstract corresponding to the first disjunct of the disjunction on the right in (1) corresponding to the null set).

The problem for the Scottish school is not that of finding some way to introduce the higher number systems, but rather that of defending the claim that, whatever way they choose, it and the more basic introduction of natural numbers through Hume's principle have distinctive *philosophical*, and especially epistemological, significance. On this philosophical front, Wright has faced objections ever since the first reviews of his book, and his school continues to this day to struggle vigorously with the most varied criticisms from the most diverse standpoints. Wright and Hale (2001) collects various technical and polemic papers on the topic.

THE MOST INFLUENTIAL of these objections is one about which nothing has been said so far: the *bad company* objection. The objection is simply that no cogent reason has been given for assuming Hume's principle, let alone whatever other abstraction principles are needed to get beyond the natural numbers as objects to the real numbers as objects, when it is known that this principle is surrounded by other abstraction principles, or laws asserting the existence of abstracts, that are untenable because they lead to contradiction.

One such untenable abstraction principle is Law V, of course; but there is another, embarrassingly close to HP. What is generally accepted as the two-place analogue of equinumerosity is *isomorphism*, defined as follows. The *field* of a two-place R is «x: $\exists y(Rxy \vee Ryx)$», which is the union of the domain «x: $\exists yRxy$» and the range «x: $\exists yRyx$». An *isomorphism* between R_1 and R_2 is a one-to-one correspondence between their fields such that whenever x_1, y_1 correspond to x_2, y_2, respectively, then we have $R_1\, x_1\, y_1$ if and only if $R_2\, x_2\, y_2$. If an isomorphism exists, we say R_1 and R_2 are *isomorphic*, and write $R_1 \cong R_2$.

The notion of isomorphism is especially important for linear orders. When the field is finite, all linear orders on it are isomorphic. But for an infinite field, like the natural numbers, many non-isomorphic linear orders are possible. Some are well-orders, like the usual order on the natural numbers, or the unusual one that places the positive ones first, in their usual order, followed by zero. Here the first order is isomorphic to an *initial segment* of the second, that is to say, to the restriction of the second to the part of the field below a certain x (namely, zero). Some are not well-orders, like the reverse of the usual order on the natural numbers. Abstracts with respect to isomorphism are called *isomorphism types*; isomorphism types of linear orders are called *order types*; order types of well-orders are called *ordinal numbers*.

The ordinal numbers are crucial in axiomatic set theory. Today each is generally identified by set theorists with the set of its predecessors. (This means, since the natural numbers are the finite ordinals, that today von Neumann's rather than Zermelo's definition of these is generally adopted: $0 = \varnothing$, $1 = \{0\}$, $2 = \{0, 1\}$, and so on.) The first transfinite ordinal, ω, is identified with the set $\{0, 1, 2, \ldots\}$ of natural numbers, and the next transfinite ordinal $\omega + 1$ with $\{0, 1, 2, \ldots, \omega\}$. Moreover, a transfinite cardinal is identified with the first ordinal having that cardinality. Thus $\aleph_0$ is the same as ω, and $\aleph_1$ the same as ω_1, the first uncountable ordinal. Needless to say, these definitions are artificial.

Originally Cantor thought of ordinals as abstracts, ω being the order type of the natural numbers in their usual order, $\omega + 1$ the order type of the natural numbers in the unnatural order with zero moved to the end. It was a minor theorem of Cantor that *every* ordinal α has a successor $\alpha + 1$. Also for Cantor ω_1 was the order type of all the countable ordinals (or order types of different well-orders of the natural numbers), ordered by calling one ordinal α less than another ordinal β if a well-order of type α is isomorphic to an initial segment of a well-order of type β. It was a major theorem of Cantor that the relation on ordinals thus defined is a well-order.

The two-place analogue of HP would read as follows, writing $\dagger R$ for the isomorphism type of R:

$$(\mathrm{H^2P})\ \dagger R = \dagger S \leftrightarrow R \cong S$$

This abstraction principle would give us isomorphism types, including order types, including ordinal numbers understood as Cantor understood them (apart from Cantor's view that abstracts are somehow produced by mental operations of human beings). And it is hard to see what could be said in favor of accepting HP as axiomatic that would not equally apply to H^2P.

But H^2P is *not* acceptable: it leads to contradiction! This was, in effect, noted even before Russell noticed his paradox, by Cesare Burali-Forti (1897/1967), who noted that the ordinal of *all* ordinals, well-ordered according to the major theorem of Cantor cited above, would be the *largest* ordinal, contrary to the minor theorem of Cantor also cited above, which implies that there is no largest ordinal. There is no real contradiction in Cantor's views here, since Cantor, to state his view a little more carefully, understood ordinals to be the order types of well-ordered *sets*, and he never assumed there was a *set* of all ordinals. The Burali-Forti paradox did come back to haunt some axiomatic set theorists after Cantor, and in particular, it is this paradox that J. B. Rosser (1942) found in the system of the first edition of Quine's *Mathematical Logic*. And now it haunts neo-Fregeanism.

For while conventional systems of axiomatic set theory do not lead to the paradox, H^2P does, as was pointed out by Hazen and Hodes and Boolos in works cited earlier. Critics of the Scottish school press the question whether one can give a principled reason—not *ad hoc* special pleading—favoring the acceptance of Hume's principle and other abstraction principles that might be wanted to go beyond the natural numbers, but not Law V or H^2P or other abstraction principles that cause trouble.

A first thought is that there is an obvious principle of rationality that may be invoked here: it is all right to accept a *consistent* abstraction principle, but not an *inconsistent* one. There are, however, two insuperable difficulties with this suggestion. The first difficulty or objection is that Boolos (1990) has shown that each of two abstraction principles may be consistent when taken separately, and yet the two together inconsistent when taken jointly. A simplification of his example (known to Boolos himself and subsequently independently rediscovered by several others) goes as follows. Consider the equivalence **R** of *almost-coextensiveness* that holds between X and Y if and only if *for all but finitely many* x we have $Xx \leftrightarrow Yx$. The corresponding abstraction principle of form (1) above might be called *almost*-V. Considering for the moment only standard models—models in which the second-order variables range over all subsets of the domain of the first-order variables—almost V has finite ones, since *all* subsets of a finite domain are almost-coextensive, but no infinite ones, by an extension of Cantor's theorem. By contrast, Hume's principle has no finite models, since there are $n + 1$ possible cardinalities, 0 through n, for a subset of domain with n elements, but has an infinite one, the Geach model. The conjunction of almost-V and Hume's principle thus has no standard models.

This is not quite the same as saying that a contradiction can be deduced from the conjunction, but by a slight modification we get an example where that is the case. Define 17-coextensiveness as follows:

$$X \equiv_{17} Y \leftrightarrow X = Y \vee (\sim\exists_{17}\, xXx \;\&\; \sim\exists_{17}\, xYx)$$

And consider the corresponding abstraction principle, Law 17-V, giving us 17-extensions $\ddagger_{17}X$. One deduces a contradiction from the conjunction of Law 17-V and Hume's principle as follows. We may assume no number 0 through 16 is a 17-extension (else replace the original 17-extensions by new ones that are the same except when the old one was a natural number, in which case the new one is the same number plus 17). Define:

$X_{-17} =$ «x: Xx & $\sim$ x is a number 0 through 16»
$X_{+17} =$ «x: $Xx \vee x$ is a number 0 through 16»
$\ddagger X = \ddagger_{17}\, X_{+17}$

Then it can be deduced that Law V holds when first-order quantifiers are restricted to x such that $U_{-17}x$ and second-order quantifiers to X such that $X \subseteq U_{-17}$.

A second difficulty or objection is that Heck (1992) has shown that there is no effective procedure for determining whether a given abstraction principle is consistent, so that the proposed principle would leave one without effective guidance as to which abstracts may be admitted and which not. Heck's proof runs as follows. For any ϕ, the formula $\phi \vee X \equiv Y$ defines an equivalence **R** that coincides with the universal equivalence (under which all concepts are equivalent) if ϕ holds, and with coextensiveness otherwise. The assumption of the existence of an abstract $\S_{\mathbf{R}}X$ with respect to **R** for every concept X, subject to the usual principle that $\S_{\mathbf{R}}X = \S_{\mathbf{R}}Y$ if and only if $\mathbf{R}XY$, will be consistent if and only if ϕ is. And Church's theorem tells us there is no effective method of determining whether an arbitrary given ϕ is consistent or not.

As Kit Fine pointed out to the author, Heck's result actually makes artificial examples like that of $\ddagger_{17}$ above superfluous. For Heck's proof shows that with any ϕ is associated an abstraction principle that is consistent if ϕ is, *and that implies* ϕ. Suppose now that both ϕ and its negation are consistent. Then the associated abstraction principles will be individually consistent and jointly inconsistent.

So much for the first thought: *only assume consistent abstraction principles*. A second thought is that what prevents us from having standard models is always having too many abstracts, meaning more than there are objects. For otherwise we can always get a standard model by taking the sets of equivalents under **R** and mapping them one-to-one into the objects. So suppose we assume the existence of abstracts only when we can prove that **R** is *non-inflating*, in the sense that there are no more sets of equivalents under **R** than there are objects.

It is not immediately obvious how to formulate this non-inflating restriction, but a usable formulation *can* be given, as follows:

(2) $$\exists R \forall X \exists x \mathbf{R} X R[x]$$

Here $R[x]$ is the section of R at x, «y: Rxy». Since (2) says every X is **R**-equivalent to the section at some object x, (2) implies there cannot be more sets of equivalents for **R** than objects.[9]

Now there are two difficulties with requiring non-inflation to be proved before the existence of abstracts for an equivalence is allowed to be assumed. The first difficulty is that requiring proof of non-inflation makes it impossible to assume HP until it has been somehow proved that there are infinitely many objects, whereas Frege's idea was to get infinitely many objects from HP. The second difficulty is one alluded to in my review of Wright, but especially developed by Kit Fine (2002), who calls it the problem of *hyperinflation*, and treats it at considerable length. The problem is that one cannot assume, even for equivalences on *objects*, and even for those that have only *two* sets of equivalents and so are about as far as possible from being inflationary, that abstracts always exist.

For the inconsistent naive theory of sets can be interpreted in a theory that allows unlimited object-abstraction, by identifying $\{x: \phi(x)\}$ and its complement with the two abstracts for the equivalence Rxy given by

$$(\phi(x) \mathbin{\&} \phi(y)) \vee (\sim\phi(x) \mathbin{\&} \sim\phi(y))$$

And a parallel difficulty arises for concept-abstraction, identifying $\ddagger X$ and its complement with the two abstracts for the equivalence **R** given by

$$(X \equiv Z \mathbin{\&} Y \equiv Z) \vee (\sim X \equiv Z \mathbin{\&} \sim Y \equiv Z)$$

Here we have a problem not so much of bad company but of *too much* company. Even if the company is good, we have too much of a good thing. To be sure, the problem only arises if it is assumed that, when dealing with more than one equivalence, the abstracts with respect to distinct equivalences are distinct, at least if the corresponding sets of equivalents are; but that is an eminently reasonable and intuitive assumption.

3.3 The General Theory of Abstraction: Its Scope

The main difference between the Scottish school and Fine is that the former are content to proceed piecemeal, adding specific abstraction principles one by one, while the latter wishes to develop a *general* theory that will, so to speak, admit all admissible abstraction principles at once. The development of such a theory is the aim of (the more technical second half of) the monograph Fine (2002), which is essentially devoted to examining thoroughly the only known simple and natural restriction that can prevent hyperinflation.

The key notion for Fine is *invariance*, which is defined as follows. A *permutation* of objects is a one-to-one correspondence P between the universal concept U and itself. In connection with permutations, we write $P(x)$ for the unique y such that Pxy. A permutation P of objects induces a permutation $\mathbf{P}$ of concepts, defined as follows:

$$\mathbf{P}XY \leftrightarrow \forall x \forall y (Pxy \rightarrow (Xx \leftrightarrow Yy))$$

By abuse of language we write $P(X)$ for $\mathbf{P}(X)$. A permutation P of objects similarly induces a permutation of relational concepts. We then define a concept X to be *P-invariant* if X is coextensive with $P(X)$ and *invariant* if it is P-invariant for every permutation P of objects. These notions can also be applied to relational concepts.

A little thought shows that there are only two invariant concepts (the universal and null concepts), and four invariant relational concepts (the universal, identity, distinctness, and null relations). In general, writing exp(n) for 2^n, the number of invariant n-place relational concepts for $n \geq 1$ is exp(exp ($n - 1$)). The proof is left as an exercise for the reader.

Things will turn out to be much more interesting one level up, so let me set down the relevant definitions. A higher-level **X** is invariant if for every permutation P of objects, $P(X)$ falls under **X** whenever X does. (Note that it is only those permutations on concepts that are induced by permutations on objects that are being considered. If one allowed arbitrary permutations of concepts, the situation would be the same as it was one level down.) A higher-level **R** is invariant if for every permutation P of objects, $P(X)$ and $P(Y)$ fall under **R** whenever X and Y do. A higher-level **R** is *doubly* invariant if for every pair of permutations P and Q of objects, $P(X)$ and $Q(Y)$ fall under **R** whenever X and Y do. These last two notions are of most interest when **R** is an equivalence. Double invariance for an equivalence amounts to the ostensibly weaker requirement that X is always **R**-equivalent to $P(X)$.

In general, a concept, relational concept, higher-level concept, or higher-level relational concept that is explicitly definable in purely logical terms (without parameters) will be invariant. There will be more invariant second-level concepts than just those explicitly definable in purely logical terms, but *all* invariant second-level concepts are regarded as "topic-neutral" and "logical" in a generalized sense by workers in such branches of mathematical logic as abstract model theory.[10] Thus requiring invariance for the equivalence involved is arguably a *principled* and not an *ad hoc* restriction on abstraction principles. It is the central restriction imposed in Fine's general theory of abstracts in order to avoid hyperinflation.

WE CAN NOW introduce a preliminary version of Fine's general theory of abstraction. It is most straightforwardly presented as a dyadic third-order theory. It contains a two-place function symbol § that takes variables of type **R** in its first place, and variables of type X in its second place, and produces a result that can be substituted for variables of type x. It thus represents an operation taking a second-level, two-place relational concept and a first-level, one-place concept as arguments to some object as value. Generally §**R**X will be written more suggestively $\S_{\mathbf{R}}X$, and called the *abstract* of X with respect to **R**. We assume extensionality, so that $\mathbf{R} \equiv \mathbf{S}$ implies $\S_{\mathbf{R}}X = \S_{\mathbf{S}}X$.

A notion of *suitability* will be defined, and the other axioms will be formulated in terms of this notion. In order to avoid trivialities, it is conventionally assumed in modern logic that there is at least one object. In the present context, it will be convenient to assume that there are at least two objects, and that we have a name for one of them, say, o. In order that $\S_{\mathbf{R}}X$ may always be defined, even when **R** is unsuitable, we conventionally let it be o in that case, and in no other. The axioms will then read as follows, where the first merely expresses the convention just adopted in order to have $\S_{\mathbf{R}}X$ always defined:

(A0) $\quad \mathbf{R}\text{ suitable} \leftrightarrow \S_{\mathbf{R}}X \neq \text{o}$

(A1) $\quad \mathbf{R}\text{ suitable} \;\&\; \mathbf{S}\text{ suitable} \;\&\; \S_{\mathbf{R}}X = \S_{\mathbf{S}}Y \rightarrow \forall Z(\mathbf{R}XZ \leftrightarrow \mathbf{S}YZ)$

(A2) $\quad \mathbf{R}\text{ suitable} \rightarrow (\S_{\mathbf{R}}X = \S_{\mathbf{R}}Y \leftrightarrow \mathbf{R}XY)$

(A1) expresses the assumption noted earlier that abstracts with respect to distinct equivalences are distinct, at least unless the corresponding sets of equivalents are the same. Two rival strengthenings suggest themselves, one being the principle that abstracts with respect to distinct equivalences are *never* the same, the other that they are the same *just* in case the corresponding sets of equivalents are. These may be expressed as follows, though neither will be adopted as an axiom for the moment:

(A1a) $\mathbf{R}$ suitable & $\mathbf{S}$ suitable & $\sim\mathbf{R} \equiv \mathbf{S} \rightarrow \S_{\mathbf{R}} X \neq \S_{\mathbf{S}} Y$

(A1b) $\mathbf{R}$ suitable & $\mathbf{S}$ suitable $\rightarrow (\S_{\mathbf{R}} X = \S_{\mathbf{S}} Y \leftrightarrow \forall Z(\mathbf{R}XZ \leftrightarrow \mathbf{S}YZ))$

(A2) expresses the usual law for abstracts, of which Law V and Hume's principle are the special cases for coextensiveness and equinumerosity. (It will turn out that in the general theory, coextensiveness is unsuitable, and equinumerosity suitable, so we have Hume's principle, which we want, and don't have Law V, which we don't.) A little thought shows that (A2) implies that any suitable **R** is an equivalence.

The weaker we make the notion of suitability, the stronger we will make the general theory of abstracts. Let us begin with a rather strong notion of suitability, and a correspondingly weak theory. Let us call **R** *strongly* suitable if the following hold:

(1) **R** is an equivalence
(2) **R** is doubly invariant
(3) **R** is a dichotomy

This last means that **R** has at most two sets of equivalents, or in other words, that there exist X and Y such that every Z is **R**-equivalent to one or the other of them. This definition gives the third-order *weak* general theory of abstractions WFTA[3].

Let me mention in passing an alternative formulation that is possible. Given any invariant **X**, it is easily seen that the following condition defines a doubly invariant dichotomy $\mathbf{R}_{\mathbf{X}}$:

$$\mathbf{R}_{\mathbf{X}} XY \leftrightarrow (\mathbf{X}X \ \& \ \mathbf{X}Y) \vee (\sim\mathbf{X}X \ \& \ \sim\mathbf{X}Y)$$

Obviously the complement of **X** would give rise to the same equivalence $\mathbf{R}_{\mathbf{X}}$. Inversely, given any doubly invariant dichotomy **R**, it is equally easily seen that the following conditions, wherein U is the universal concept, define complementary invariant second-level concepts:

$$\mathbf{X_R}^+ X \leftrightarrow \mathbf{R}\mathrm{U}X \qquad \mathbf{X_R}^-(X) \leftrightarrow {\sim}\mathbf{R}\mathrm{U}X$$

Assuming the existence of abstracts for doubly invariant dichotomies is in effect equivalent to assuming the existence of extensions for invariant second-level concepts.

Now invariant second-level concepts amount to what in abstract model theory are called *generalized quantifiers*. These include, for instance: *none*, *some*, *all*, *at most one*, *exactly one*, *at least one*, *at most two*, *exactly two*, *at least two*, *finitely many*, *infinitely many*, *most*, *evenly many*, *oddly many*, and so on. The assumption of the existence of logical *objects* corresponding to these quantifiers at once gives us surrogates for natural numbers, in the form of the objects corresponding to *exactly zero*, *exactly one*, *exactly two*, and so on. Recognizing these objects corresponds to the historical step of switching from the adjectival use of numerals as in "Four horses pulled the chariot," to the nominal use of numerals as in "Four is a perfect square." Having surrogates for numbers among our objects, concepts under which they fall give us surrogates for sets of natural numbers, and we can therefore interpret second-order Peano arithmetic P^2 in our theory.

But actually we can do better, since the quantifier-objects themselves already include surrogates for sets of natural numbers. For instance, the objects corresponding to *evenly many* and *oddly many* can serve as surrogates for the sets of even and of odd numbers. Such is the idea of Fine's interpretation of analysis with real numbers as *objects* in his theory; a somewhat more rigorous treatment will be given shortly.

There is also a second-order version WFTA^2. In this theory we have a rule permitting us, for any formula $\theta(X, Y)$, to introduce a new symbol § and assume (A0)–(A2) with θ in place of

R.[11] The formula θ may contain parameters, that is, it may look like $\theta(\boldsymbol{U}, \boldsymbol{u}, X, Y)$ in which case § will have extra argument places for these parameters, and we will write $\S_{\boldsymbol{Uu}}X$ for $\S X\boldsymbol{Uu}$. Corresponding to extensionality in the third-order case we assume for formulas θ and χ and associated symbols § and ¶, the following:

$$\forall X \forall Y(\theta(\boldsymbol{U}, \boldsymbol{u}, X, Y) \leftrightarrow \chi(\boldsymbol{V}, \boldsymbol{v}, X, Y)) \rightarrow \S_{\boldsymbol{Uu}}X = \P_{\boldsymbol{Vv}}Y$$

A rigorous treatment of the interpretation of analysis shows that P^3 is interpretable in $WFTA^2$ using only *two* special symbols. We first note that for any U the following formula $\phi(U, X, Y)$ determines an equivalence:

$$(4) \qquad (X \approx U \,\&\, Y \approx U) \vee (\sim X \approx U \,\&\, \sim Y \approx U)$$

Note that for any permutation P and any X we have $P(X) \approx X$. Then a little thought shows that $\phi(U, X, Y)$ determines a *doubly invariant* equivalence—and the same one for two different parameters U_1 and U_2 if and only if $U_1 \approx U_2$. Obviously it is a dichotomy. So $\phi(U, X, Y)$ is suitable, and we may introduce a symbol † and assume the following:

$$(5) \quad \dagger_U X = \dagger_U Y \leftrightarrow (X \approx U \,\&\, Y \approx U) \vee (\sim X \approx U \,\&\, \sim Y \approx U)$$

A little thought shows that defining $\#U = \dagger_U U$ gives us Hume's principle.

We next note that for any U the following formula $\psi(U, X, Y)$ determines an equivalence:

$$(6) \qquad (U(\#X) \,\&\, U(\#Y)) \vee (\sim U(\#X) \,\&\, \sim U(\#Y))$$

Now for any permutation P and any X we have $\#P(X) = \#X$, and this fact shows that ψ determines a *doubly invariant* equivalence. Moreover, inspection of (6) shows that it is a dichotomy.

So ψ is suitable, and we may introduce a symbol ¶ and assume the following:

$$(7)\quad ¶_U X = ¶_U Y \leftrightarrow (U(\#X) \,\&\, U(\#Y)) \vee ({\sim}\, U(\#X) \,\&\, {\sim}\, U(\# Y))$$

A little thought shows that (7) implies that for any U there exists a unique u for which the following holds:

$$(8)\qquad ({\sim}\exists XU(\#X) \,\&\, u = 0) \vee (\exists XU(\#X) \,\&\, u = ¶_U X)$$

and letting $‡_N U$ be this u, this gives us numerical Law V. This argument, essentially Fine's with a bit of bookkeeping added, establishes the following result:

> ***Refined Frege-Fine Theorem.*** For all $n \geq 2$, $(n + 1)$st-order Peano arithmetic can be interpreted in the nth-order weak general theory of abstraction.

3.4 The General Theory of Abstraction: Its Limits

It is possible to develop a stronger general theory of abstraction by adopting a weaker notion of suitability. Let us change the definition so that we now consider **R** *weakly* suitable if and only if we have the following:

(1) **R** is an equivalence
(2) **R** is very nearly invariant
(3) **R** is non-inflating

The notion of non-inflating in (3) has already been discussed.

The notion in (2) needs to be defined. Call *W small* if *W* is not equinumerous with the universal concept U, and call *W very small* if there is some small *W'* such that *W* is equinumerous with a proper part of *W'*, but not with all of *W'*. We say a permutation P of objects fixes W if $P(w) = w$ whenever Ww, and we say that a concept, relational concept, higher-level concept, or higher-level relational concept is *W-invariant* if it is P-invariant for every permutation P of objects that fixes W. Finally, we say *nearly invariant* to mean W-invariant for some small W, and *very nearly invariant* to mean W-invariant for some very small W. Thus, unpacking the definitions, $\mathbf{R}$ is very nearly invariant if there is some very small W such that for any permutation P for which $P(w) = w$ whenever Ww, and for any X and Y, we have $\mathbf{R}XY$ if and only if $\mathbf{R}P(X)P(Y)$. With the weakened definition (1)–(3) of suitability we get the *strong* general theory of abstraction $SFTA^3$; there is also a second-order variant $SFTA^2$.

Fine devotes considerable space to a thorough study of standard models of $SFTA^3$ with domains of objects of various transfinite cardinalities, considering both how such models can be defined "from the top down" and how they can be constructed "from the bottom up." (Indeed, the part of the subject treated in Fine's work is the only part of the subject of consistent modifications of Frege's inconsistent system for which a really thorough and systematic examination of models has been undertaken.) For purposes of proving the consistency of the theory—the only question to be taken up here—the existence of just one model, which need not be standard, is of course enough. And so for present purposes it will be enough to extract and condense just enough out of Fine's extended discussion to give us one model, adding some bookkeeping to turn the model construction into a proof of interpretability, a topic Fine does not directly address. Unfortunately, from this point on it will be necessary to assume more familiarity with axiomatic set theory than I have been assuming so far, though the information that can be obtained from the earlier parts of any introductory textbook should be more than enough.

In constructing a model to prove consistency for the strong general theory of abstraction, I will also assume the axiom of constructibility $V = L$. All that one needs to know about this axiom is that it implies the continuum hypothesis CH, and implies that a certain relation $<_L$ definable by a certain formula $\lambda(x, y)$ of the language of set theory gives a well-order of all sets. This feature can be used whenever we need to make a choice of a set with a certain property and want to be explicit about which choice we are making: we then choose the $<_L$-least set with the property. But this feature will be used only at the very end of the proof. The assumption of V = L is harmless. For Gödel showed that ZF + V = L is a conservative extension of ZF for arithmetical statements, which is to say that any conclusion that could be stated in the language of first-order arithmetic that is proved with the aid of this hypothesis could be proved without it; and consistency statements can be coded in the language of first-order arithmetic (as they are in textbooks in connection with the second incompleteness theorem).

IN THE MODEL to be constructed, the domain of the object variables will be ω_1, the first uncountable ordinal, identified with the set of all countable ordinals. The model will be standard, so the domain of the one-place concept variables will be the power set $\wp(\omega_1)$ or set of all subsets of ω_1, and the domain of the two-place relational concept variables will be the power set $\wp(\omega_1 \otimes \omega_1)$ of the Cartesian product $\omega_1 \otimes \omega_1$ or set of all ordered pairs of elements of ω_1, that is, the set of all two-place relations on ω_1. In the third-order version of the theory, the second-level one-place concept variables and two-place relational concept variables will range over $\wp(\wp(\omega_1))$ and $\wp(\wp(\omega_1) \otimes \wp(\omega_1))$. Since the model is standard, all comprehension axioms of higher-order logic will hold, as will extensionality.

It remains to consider the axioms about abstracts. What we need to do to get a model is to associate to any suitable equivalence $\mathbf{R}$ and any $X \subseteq \omega_1$ an ordinal $\S_{\mathbf{R}}X < \omega_1$ in such a way that we have:

(4) $$\mathbf{R}XY \leftrightarrow \S_{\mathbf{R}}X = \S_{\mathbf{R}}Y$$

(5) $$\S_{\mathbf{R}}X = \S_{\mathbf{S}}Y \rightarrow \forall Z(\mathbf{R}XZ \leftrightarrow \mathbf{S}YZ)$$

The construction can actually be so arranged as to give whichever of the following incompatible strengthenings of (5) one prefers:

(5a) $\mathbf{R} \neq \mathbf{S} \rightarrow \S_{\mathbf{R}} X \neq \S_{\mathbf{S}} Y$

(5b) $\S_{\mathbf{R}} X = \S_{\mathbf{S}} Y \leftrightarrow \forall Z(\mathbf{R}XZ \leftrightarrow \mathbf{S}YZ)$

For the construction we will, of course, be using the suitability hypothesis on the equivalence **R**. Let us first consider what "very nearly invariant" amounts to in our present context. Since the domain of the object variables has size $\aleph_1$, "small" will mean $< \aleph_1$, which is to say $\leq \aleph_0$, or countable, and "very small" will mean $< \aleph_0$, or finite. So if **R** is very nearly invariant, there will be a finite subset s of ω_1 such that **R** is s-invariant. The definition of "non-inflating" was a technical condition sufficient (and actually, necessary, though we will not need that fact) to imply having $\leq \aleph_1$ sets of equivalents.

What is crucial to the construction is to establish that there are only $\aleph_1$ pairs consisting of a suitable equivalence **R** and one of its sets of equivalents $\mathbf{R}[X]$. The non-inflating condition tells us there will not be too many sets of equivalents for any given suitable equivalence, so what is crucial will be to establish that there are not too many very nearly invariant equivalences. The key to doing so will be locating a condition that for any finite $s \subseteq \omega_1$ and $X_1, Y_1, X_2, Y_2 \subseteq \omega_1$ will be sufficient to imply that there exists a permutation P of ω_1 that fixes s and has $P(X_1) = X_2$ and $P(Y_1) = Y_2$.

The key to locating such a condition will be the following technical notion. Call p a *profile* if p is an octuple whose first four components are finite subsets of ω_1 and whose last four components are cardinals $\leq \aleph_1$, and call p an s-profile if the union of its first four components is s. For given X and Y, define *their* s-profile $\pi_s(X, Y)$ to be the one with the following eight components:

$$\begin{array}{cccc} (X \cap Y) \cap s & (X - Y) \cap s & (Y - X) \cap s & (\omega_1 - X - Y) \cap s \\ \|(X \cap Y) - s\| & \|(X - Y) - s\| & \|(Y - X) - s\| & \|(\omega_1 - X - Y) - s\| \end{array}$$

Then I claim that if $\pi_s(X_1, Y_1) = \pi_s(X_2, Y_2)$, then there is a permutation P of ω_1 that fixes s and has $P(X_1) = X_2$ and $P(Y_1) = Y_2$. To prove this claim, begin by noting that the sameness of the first four components of the s-profiles implies that the identity I_s satisfies $I_s(X_1 \cap s) = X_2 \cap s$ and $I_s(Y_1 \cap s) = Y_2 \cap s$. To obtain the desired P it will be enough to find a permutation Q of $\omega_1 - s$ with $Q(X_1 - s) = X_2 - s$ and $Q(Y_1 - s) = Y_2 - s$, for then I_s and Q can be combined in the obvious way to give P. And the desired Q can be obtained by combining four one-to-one correspondences, one between $(X_1 \cap Y_1) - s$ and $(X_2 \cap Y_2) - s$, another between $(X_1 - Y_1) - s$ and $(X_2 - Y_2) - s$, another between $(Y_1 - X_1) - s$ and $(Y_2 - X_2) - s$, and another between $(\omega_1 - X_1 - Y_1) - s$ and $(\omega_1 - X_2 - Y_2) - s$. These four exist by the sameness of the last four components of the s-profiles.

Now call a set S *fair* if for some finite $s \subseteq \omega_1$, S is a nonempty set of s-profiles. The s in question will be unique and can be recovered from S as the union of the first four components of any profile p in S. Let us define the S-profile $\pi_S(X, Y)$ to be simply $\pi_s(X, Y)$ for this s. And let us define X and Y to be S-related if $\pi_S(X, Y) \in S$. I claim that every very nearly invariant equivalence **R** coincides with S-relatedness for some fair S. To prove the claim, let **R** be given. Then since **R** is very nearly invariant, there is a finite $s \subseteq \omega_1$ such that **R** is s-invariant, meaning that for any permutation P fixing s we have $\mathbf{R}XY$ if and only if $\mathbf{R}P(X)P(Y)$ for all X and Y. Let S be the set of s-profiles $\pi_s(X, Y)$ for X and Y such that $\mathbf{R}XY$. Reviewing the definitions, it is immediate that $\mathbf{R}XY$ implies that X and Y are S-related. To complete the proof of the claim it will be enough to show that the converse holds. So suppose X and Y are S-related, meaning that $\pi_s(X, Y) = \pi_S(X, Y) \in S$, and therefore by the definition of S implying that $\pi_s(X, Y) = \pi_s(X_1, Y_1)$ for some X_1 and Y_1 such that $\mathbf{R}X_1\, Y_1$. But since $\pi_s(X, Y) = \pi_s(X_1, Y_1)$, there exists a permutation P that fixes s and has $P(X_1) = X$ and $P(Y1) = Y$, from which it follows by the s-invariance of **R** that we have **R***XY* as required to complete the proof of the claim.

Call a set *S* *good* if *S* is fair and *S*-relatedness is a suitable equivalence. Then what has just been shown is that for every suitable equivalence **R** there is a good *S* such that **R** coincides with *S*-relatedness. And what we have accomplished so far is to show that there are no more suitable equivalences **R** than there are good *S*. We need next a bound on the number of such *S*. I claim there are only $\aleph_1$ fair *S* and *a fortiori* only $\aleph_1$ good *S*.

Since each fair *S* is a set of *s*-profiles for some finite $s \subseteq \omega_1$, and there are only $\aleph_1$ such *s*, to prove the claim it will be enough to show that for any given *s* the number of sets *S* of *s*-profiles is no more than $\aleph_1$. Now every such *S* is a subset of the set of *all* *s*-profiles *p*, and the set of all *s*-profiles is countable. For each of the first four components of such a *p* is a subset of *s*, and there are only finitely many of those, while each of the last four components of such a *p* is a cardinal $\leq \aleph_1$, and there are only countably many of those, namely the natural numbers and $\aleph_0$ and $\aleph_1$. To complete the proof of the bound on the number of very good *S*, we invoke for the first and last time CH, which tells us that the number of subsets of a countable set is $\aleph_1$.

To complete the proof of consistency, it would be enough now to assign each pair ⟨**R**, **X**⟩ consisting of a suitable equivalence **R** and one of its sets of equivalents **X** a distinct ordinal < ω_1, which we now know is possible, since we have seen that there are only $\aleph_1$ suitable equivalences, and there are at most $\aleph_1$ sets of equivalents for each, making only $\aleph_1$ such pairs in all. This done, we could take $\S_{\mathbf{R}}X$ to be the ordinal assigned to ⟨**R**, **R**[*X*]⟩, and we would have a model of (4) and (5a). A slightly modified construction, assigning ordinals only to the sets of equivalents **X** rather than pairs ⟨**R**, **X**⟩, would give (5b) instead. We would have consistency for either option, and without using more of V = L than its implication CH.

THOUGH INDEED this *is* enough to prove consistency, it will be worthwhile to make use of $<_L$ and give a more explicit definition of $§_{\mathbf{R}}X$. To begin with, call S *very good* if S is good and there is no $T <_L S$ such that T is good and T-relatedness coincides with S-relatedness. Then every suitable $\mathbf{R}$ coincides with S-relatedness for some *unique* very good S. Call a pair $\langle S, Y\rangle$ *excellent* if S is very good, $Y \subseteq \omega_1$, and Y is the $<_L$-least member of its set of equivalents under S-relatedness, or in other words, there is no $Z <_L Y$ such that Y and Z are S-related. Since S-relatedness is a non-inflating equivalence, there are at most $\aleph_1$ sets of equivalents, each of which has exactly one $<_L$-least member, so that in all there are $\leq\aleph_1$ excellent pairs $\langle S, Y\rangle$ for any one of the $\aleph_1$ very good S, and hence only $\aleph_1$ excellent pairs altogether. Moreover, for any pair $\langle \mathbf{R}, X\rangle$ consisting of a suitable equivalence $\mathbf{R}$ and a subset of ω_1, there is a unique excellent pair $\langle S, Y\rangle$ such that $\mathbf{R}$-equivalence coincides with S-relatedness, and X is S-related to Y, which is to say, X is $\mathbf{R}$-equivalent to Y. (We already know there is a unique very good S such that $\mathbf{R}$-equivalence coincides with S-relatedness, and it is also clearly the case that there is a unique Y that is $<_L$-least among all subsets of ω_1 that are $\mathbf{R}$-equivalent to X.) Let us say that such a pair *captures* $\mathbf{R}$ and X. Since the number of excellent pairs is $\aleph_1$, there is a one-to-one correspondence between ordinals $< \omega_1$ and excellent pairs, and among all such correspondences there is one that is $<_L$-least, which we may call the canonical correspondence. Then for any pair $\langle \mathbf{R}, X\rangle$ consisting of a suitable equivalence $\mathbf{R}$ and a subset of ω_1, there is a unique ordinal $x < \omega_1$ such that x corresponds under the canonical correspondence to the excellent pair that captures $\mathbf{R}$ and X. Let $\sigma(\mathbf{R}, X)$ be this unique ordinal. Reviewing the definitions, we see that if $\mathbf{R}$ and $\mathbf{S}$ do not coincide, then $\sigma(\mathbf{R}, X) \neq \sigma(\mathbf{S}, Y)$ for any X and Y, while for fixed $\mathbf{R}$ we have $\sigma(\mathbf{R}, X) = \sigma(\mathbf{R}, Y)$ if and only if X and Y are $\mathbf{R}$-equivalent. Therefore if we take $§_{\mathbf{R}}\, X$ to denote $\sigma(\mathbf{R}, Y)$, we get a model of (4) and (5a). A slightly modified construction would give (5b) instead.

This *really* is enough to complete the proof of consistency, but looked at closely, the proof actually proves *more* than consistency. The definition of a model in effect gives an interpretation of SFTA[3] in ZFC + V = L. Moreover, though the proof used the existence of ω_1 and $\wp(\omega_1)$ to get domains for the object- and first-level concept variables and thereby used the existence of $\wp(\omega)$ and $\wp^2(\omega)$, and though the proof used subsets of $\wp(\omega_1)$ as the domain of the second-level concept variables, the interpretation nowhere really uses the assumption that these subsets form a set, which is to say that we do not need $\wp^2(\omega_1)$ or $\wp^3(\omega)$ for the proof. This means that we actually get an interpretation in $\mathrm{ZF}^- + \wp^2(\omega) + \mathrm{V} = \mathrm{L}$. And in our preliminary survey of systems of analysis and set theory it was remarked that this theory is interpretable in $\mathrm{ZF}^- + \wp^2(\omega)$, which in turn is interpretable in P^4. This gives the case $n = 3$ in the following theorem, from which the cases for $n > 3$ follow trivially:

> ***Converse to the Frege-Fine Theorem.*** For any $n \geq 2$, the nth-order strong general theory of abstraction is interpretable in $(n + 1)$st-order Peano arithmetic.

The case $n = 2$ is a little more delicate, and I will not dot every *i* or cross every *t*. For the cognoscenti, what the construction (in its more explicit version, using $<_{\mathrm{L}}$) *really* shows is that there is a way to associate to any formula $\theta(\boldsymbol{U}, \boldsymbol{u}, X, Y)$ of the language of set theory a formula $\theta^*(\boldsymbol{U}, \boldsymbol{u}, X, x)$ of the language of set theory such that for any $\boldsymbol{U}$ and $\boldsymbol{u}$, if $\theta(\boldsymbol{U}, \boldsymbol{u}, X, Y)$ defines a suitable equivalence on subsets of ω_1, then $\theta^*(\boldsymbol{U}, \boldsymbol{u}, X, x)$ defines an assignment of abstracts $x < \omega_1$ to subsets $X \subseteq \omega_1$ such that (4) and (5a) hold with $\theta(\boldsymbol{U}, \boldsymbol{u}, X, Y)$ as $\mathbf{R}XY$ and the unique x such that $\theta^*(\boldsymbol{U}, \boldsymbol{u}, X, x)$ as $\S_{\mathbf{R}} X$. The formula θ^* says that, letting σ be the canonical enumeration of excellent pairs and $\langle S, Y\rangle = \sigma(x)$, then the pair $\langle S, Y\rangle$ captures the pair consisting of the relation defined by $\theta(\boldsymbol{U}, \boldsymbol{u}, X, Y)$ together with X. That the relation defined by $\theta(\boldsymbol{U}, \boldsymbol{u}, X, Y)$ exists as a *set* **R** is not really required. And thus $\wp(\omega_1)$ and $\wp^2(w)$ are not really required, and we get an interpretation in $ZF^- + \wp(\omega) + V = L$, and thence in $ZF^- + \wp(\omega)$, and thence in P^3.

The final result is that, if we stay within the bounds of second-order logic as most recent workers do, then relying on invariance to solve the hyperinflation problem, the "limits of abstraction" turn out to be those of third-order Peano arithmetic. This is really a quite strong theory, and any mathematics that cannot be done in it might well be called "higher set theory," so that Fine's approach can be said to get us all of mathematics except higher set theory.

Unfortunately, however, Fine's approach does *not* give a complete answer to the bad-company objection, as Fine himself in effect concedes. The problem is that for any R, the relation defined as follows:

$$\mathbf{E}_R ST \leftrightarrow (S \cong R \;\&\; T \cong R) \vee ({\sim}\, S \cong R \;\&\; {\sim}\, T \cong R)$$

is invariant and a dichotomy; and yet to assume abstracts exist for all such equivalences is to assume the existence of isomorphism types, including order types, including ordinal numbers, and the Burali-Forti paradox ensues. An appropriate extension of Fine's approach from equivalences on one-place concepts to equivalences on two-place relational concepts has not (yet) been developed.

3.5 Frege-Inspired Set Theories

The common feature of the Wright-Hale and Fine abstractionist approaches is that they dispense with Law V and extensions, and work with abstracts for equivalences that are quite different from coextensiveness.[12] I wish now to turn to theories that instead save Frege from contradiction by restricting Law V. Such theories keep extensions, though subject to a restriction, and like Frege's original theory they do not directly assume abstracts for any equivalence other than coextensiveness.

In a sense, of course, *any* axiomatic set theory can be viewed as a restriction on Frege's inconsistent set theory. The set theories I especially wish to examine, however, have two features that make them similar to Frege's original theory and different from mainstream axiomatic set theories such as ZFC. The first such feature is that they make serious use of second-order logic, and that in accordance with Frege's principles they subordinate the notions of *set* and *element* to those of *falling-under* and *extension-of*. The second such feature is that they have only a single, unified set-existence axiom, or at most a couple, in contrast to conventional axiomatizations that introduce sets in driblets through a series of a half-dozen or more existence axioms.

To begin with, let me set up a common framework for a variety of such set theories. We have full second-order logic, and, to begin with at least, I will assume that we have full *dyadic* second-order logic. And we have special relation-symbols € and ß and ∈ with the following axioms. First, we have the *axioms of subordination*, defining set and element in terms of falling-under and extensions-of:

(1) $\text{ß}x \leftrightarrow \exists X\, \text{€}xX$

(2) $x \in y \leftrightarrow \exists Y(\text{€}yY \,\&\, Yx)$

Second, we have the *axiom of extensionality* in the following Fregean form:

(3) $\text{€}xX \,\&\, \text{€}yY \rightarrow (x = y \leftrightarrow X \equiv Y)$

Note that (3) implies that the X in (1) and the Y in (2) are unique. We have seen how (3) implies the uniqueness of the extension when it exists, permitting the definition of ‡X and {x: $\phi(x)$} for those concepts X and «x: $\phi(x)$» that have extensions. And we have seen how (1)–(3) permit the deduction, among other things, of the following alternate formulation of extensionality:

(4) ßx & ßy & $\forall z(z \in x \leftrightarrow z \in y) \rightarrow x = y$

Then we have a set-existence axiom of one of the two forms:

(5a) $\exists x$ €$xX \leftrightarrow X$ safe

(5b) X safe $\rightarrow \exists x$ €xX

I will concentrate on those with axioms in biconditional form (5a). We have seen how such a restricted existence assumption could be artificially presented as the assumption of the existence of abstracts for a certain equivalence (agreeing with ≡ for safe X but counting all unsafe X as equivalent), and that adopting this artificial presentation is equivalent to assuming one additional nameable object o, not the extension of any concept.

One type of set theory fitting this format we have considered already, the *zig-zag* theories where safety is understood as a kind of simplicity. In this section we will have a look at other types. One such is discussed in Fine's book. The suggestion is that we take "safe" to mean "nearly invariant," so the existence axiom reads as follows:

(6) $\exists x$ €$xX \leftrightarrow X$ nearly invariant

A little thought shows that if X is nearly invariant, so that there is some small W such that X is W-invariant, then either all x outside W have $\sim Xx$ or all x outside W have Xx. This means that either X itself is small, or its complement is small, in which case we say X is *co-small*. It is easily seen that conversely any set that is small or co-small is nearly invariant. Thus the set-existence axiom (6) is equivalent to the following:

$$\exists x \in xX \leftrightarrow X \text{ small} \vee X \text{ co-small} \tag{7}$$

As Stewart Shapiro reminds me, the set theory based on this variant axiom (7) was (briefly) considered at the end of Boolos (1993). There Boolos says that it, and the question of its consistency, was first brought to his attention by T. Parsons. The Fregean set theory with the axiom (6) (or equivalently (7)), I will accordingly call *Parsons set theory*.

It is interpretable in second-order Peano arithmetic. For in P^2 small amounts to finite, and there are simple codings of finite sets of natural numbers that are definable in P^2 (and indeed already in P^1 and indeed already in still weaker theories), and these can easily be extended to give a coding of finite and co-finite sets. (One coding of finite sets codes s by the sum $\sigma(s)$ of the powers 2^m of 2 for $m \in s$, and this can be modified to a coding in which $2 \cdot \sigma(s)$ codes s and $2 \cdot \sigma(s) + 1$ codes the complement of s.) And this is all the machinery needed to give an interpretation.

Conversely, working in Parsons set theory, since there is at least one object, an empty concept is small, and hence the empty set Ø exists, and the universal concept is co-small, and hence the universal set V exists. Moreover, since there is at least one object, $V \neq Ø$, and there are in fact at least two objects. This is enough to give us singletons and co-singletons, and in particular the axioms of UUST, which we know to be enough, with dyadic second-order logic, to permit the interpretation of P^2.

All this was known to Fine and Boolos. Let me set it down as an official theorem, with a name that will distinguish it from those authors' larger theorems.

The Little Fine-Boolos Theorem. Parsons set theory and second-order Peano arithmetic are interpretable in each other.

The remaining type of set theory to be considered is based on an idea already mentioned, that of *limitation of size*, whose role in the history of set theory from Cantor to the present day has been examined in detail by Michael Hallett (1984). I will distinguish four forms of the idea, two *quantitative* and two *non-quantitative*, and two *maximalist* and two *non-maximalist*. In all forms the idea is that X has an extension if and only if there are not too many objects falling under X. The quantitative versions are committed to understanding "many" and "few" in terms of one-to-one correspondences, while the non-quantitative are not. The maximalist versions are committed to equating "too many" with "as many as there are objects altogether," the non-maximalist versions are not.

The weakest of the four, the non-quantitative, non-maximalist version, is already sufficient to motivate Zermelo's axiom of separation:

(8) $$X \subseteq Y \to (\exists y\, \text{€}yY \to \exists x\, \text{€}xX)$$

If X is a part of Y and there are not too many objects falling under Y, then there are not too many objects falling under X. Since every X is contained in the universal concept, (8) is incompatible with the assumption that the universal concept has an extension, which is to say, with the assumption that there exists a universal set. A little thought shows that (8) is equivalent to the more conventional second-order single-axiom formulation, which with comprehension implies each instance of the axiom-scheme version:

(8a) $$\text{ß}\, y \to \exists\, x(\text{ß}x \;\&\; \forall z(z \in x \leftrightarrow z \in y \;\&\; Xz))$$

(8b) $$\text{ß}\, y \to \exists\, x(\text{ß}x \;\&\; \forall z(z \in x \leftrightarrow z \in y \;\&\; \phi(z)))$$

The quantitative non-maximalist version is sufficient to motivate Frankel's axiom of replacement, which can be stated using dyadic second-order logic thus:

(9) $$\forall y \exists!\, xRyx \;\&\; \forall x(Xx \leftrightarrow \exists y(Yy \;\&\; Ryx)) \to (\exists y\, \text{€}yY \to \exists x\, \text{€}xX)$$

Again a little thought shows that (9) is equivalent to a more conventional second-order single-axiom formulation, which in turn implies each instance of an axiom-scheme version. The details will not matter here.

The non-quantitative, maximalist version of the limitation of size idea will be of no immediate concern. The quantitative, maximalist version motivates *von Neumann's axiom*, which in the present context may be formulated as follows:

$$\exists x \, \text{€} x X \leftrightarrow X \text{ small} \tag{10}$$

Here "small" is defined in dyadic second-order logic as it was earlier, to mean "not in one-to-one correspondence with the universe of all objects." Boolos considered in Boolos (1989) and elsewhere the Fregean set theory with (10) as its set-existence axiom, or rather the abstractionist variant, where an axiom Boolos calls "New V" supplies extensions for small X plus one additional object o distinct from all extensions. Boolos used this addition *only* to get the existence of at least two objects, so let me call the set theory with (10) and $\exists_2 x(x = x)$ as existence axioms *Boolos set theory*. Then by the same reasoning as with Parsons set theory above we have the following:

> ***The Little Boolos Theorem.*** Boolos set theory and second-order Peano arithmetic are interpretable in each other.

Von Neumann's axiomatization of set theory used *predicative* dyadic second-order logic, but otherwise was Boolos set theory with a few additional set-existence axioms. With an ingenious technical improvement by Levy, the number of additional axioms needed to get full-strength set theory can be reduced to two: infinity and power set. Adding infinity gives a theory of the same strength as P^3, and adding also power set gives a theory of the same strength as ZFC. But we do need those two further axioms. For more on Boolos set theory in comparison with standard set theories, see Shapiro and Weir (1999).

The long and short of it is that the versions of the limitation of size idea considered so far do not get one very far into the realm of set theory beyond analysis. The problem is that the principle that a set exists unless its elements would be as many as there are objects does not tell us that a set exists unless it is very, very large, without the further assumption that there are very, very many objects. To get more, one has to appeal repeatedly to fresh intuitions about how many is too many: for the axiom of infinity, to the intuition that infinitely many is not too many; for the power set axiom, to an intuition implying that uncountably many is not too many; and if one wants inaccessible cardinals, to an intuition that inaccessibly many is not too many. The set theorist who relies on the versions of the limitation of size idea considered so far has to keep coming back to ask for more, like the fisherman in the fairy tale about the magic flounder. (The fisherman finally asked for too much, as did set theorists when they asked for Reinhardt cardinals.) If one is going to get out of the limitation of size idea a single, unified formal set-existence axiom sufficient to give a really powerful set theory, one is going to have to adopt a different method.

3.6 The Reflection Principle

All the theories considered so far in this chapter have made use of *dyadic* second-order logic, and the notion of equinumerosity. Let us now make a clean break, and henceforth stick with *monadic* second-order logic. In particular, we give up any quantitative version of the limitation of size idea.

What is left of our Fregean framework for set theory is monadic second-order logic with symbols € and ß and $\in$ and with the axioms of subordination and extensionality:

(1) $$ß x \leftrightarrow \exists X\, € x X$$

(2) $$x \in y \leftrightarrow \exists Y(€yY \;\&\; Yx)$$

(3) $$€xX \;\&\; €yY \rightarrow (x = y \leftrightarrow X \equiv Y)$$

What is also left is the non-quantitative version of the limitation of size idea, which even in its non-maximalist form gives the axiom of separation:

(4) $$X \subseteq Y \rightarrow (\exists y\, €yY \rightarrow \exists\, x\, €xX)$$

What is missing is a further set-existence axiom of the form

(5) $\sim\exists x\, €xX \rightarrow$ there are too many x falling under X to form a set

Now one concept that we know cannot have an extension is the universal concept. Applied to this case, (5) tells us this:

(6) there are too many objects to form a set

The weakness of the von Neumann approach was that it in effect *defined* "too many to form a set" to mean "as many as there are objects," thus depriving (6) of any serious substance. What we need is some way of understanding "too many to form a set" that will leave (6) with some usable content.

Now there is a word that appears in this connection in writers as diverse as Dummett (1991) and Cantor (1885), the word "indefinite" (or "*unbestimmt*"). There is no set of all objects, according to both thinkers, because there are indefinitely many objects. Dummett, who uses the expression "indefinitely extensible," is soon led by his intuitionistic proclivities to understand indefiniteness as being of a matter of some sentences lacking truth-values. Cantor, who uses the expression "quantitatively indefinite" (or "*quantitativ unbestimmt*"), understands indefiniteness in the sense of being incapable of being assigned one of his alephs as cardinality. I propose to understand "indefiniteness" in a much more immediate way, as "undefinability," thus:

(7) there are undefinably many objects

And what does *that* mean? Well, it cannot mean that it is impossible to make a true statement about how many objects there are, since it is a true statement that there is at least one object, for instance. What I take (7) to mean is that though it is possible to make a true statement about how many objects there are, there are too many objects for it to be possible for such a statement to be *definitive* of how many there are: there will necessarily be not merely as many as is said, but more also. Whatever one says, *that* could still be true if it were said not about *all* objects, but about just *some* objects, fewer than all.

(8) anything that is true when said about all objects
remains true when said about just some objects, fewer than all

This looks like a promising principle, for one can imagine applying it as follows. To begin with, there is at least one object. By (8), that would still be true if one were speaking not of all objects, but just of some objects, fewer than all; in other words, it is an understatement, which means that there must be at least two objects. Then by (8) again, *that* is an understatement, so there are at least three objects. Continuing in this way, there are infinitely many objects. But by (8), even that is an understatement, so there must be uncountably many objects. But by (8) again, even *that* is an understatement, so. . . . But already we have got further than we ever got before.

And what is meant by "fewer than all" here? Having discarded dyadic second-order logic and quantitative versions of limitation of size, we still have the so far unused non-quantitative but maximalist version, according to which "fewer than all" becomes "few enough to form a set". This transforms (8) into the following:

(9) anything that is true when said about all objects remains true when said about just the elements of some set

At this point, however, paradox may seem to threaten, since one thing that is true about all objects is that they are not few enough to form a set, and *that* certainly isn't true of the elements of any set t. But the paradox can be resolved by being careful about "about". The statement "There is no set that has all objects as elements" is true, and the statement "There is no set that has all elements of t as elements" is false, but the false statement isn't *just* about the elements of t, since the universal negative "there is no set . . . " says something about *all* sets, not just those that are elements of t. The statement in the vicinity that *is* just about the elements of t is rather the statement "There is no set *among the elements of* t that has all elements of t as elements," and that is not paradoxical.

Clearly the notion required here is that of relativization of quantifiers. Our general notation for relativizing the quantifiers in a formula ϕ to a formula $\theta(x)$, replacing $\forall x(\ldots)$ by $\forall x(\theta(x) \rightarrow \ldots)$, and similarly for $\exists$, has been ϕ^{θ}. When $\theta(x)$ is $x \in t$, so we are replacing $\forall x(\ldots)$ by $\forall x(x \in t \rightarrow \ldots)$, and similarly for $\exists$, we just write ϕ^{t}. Since we are using *second*-order logic, we need to say how second-order quantifiers are to be relativized. Relativizing, $\forall X$ becomes

$$\forall X(\forall x(Xx \rightarrow \theta(x)) \rightarrow \ldots) \qquad \text{or}$$
$$\forall X(\forall x(Xx \rightarrow x \in t) \rightarrow \ldots)$$

in the general or the special case, respectively, and similarly for $\exists$. The form on the right may be abbreviated $\forall X(X \subseteq t \rightarrow \ldots)$ or even just $\forall X \subseteq t(\ldots)$. Having this notion of relativization in hand, we can say that what is meant by (9) is the following:

$$\phi \rightarrow \exists t\, \phi^{t} \tag{10}$$

It is not claimed that (10) is self-evidently true, or even self-evidently consistent. By Gödel's second incompleteness theorem, a self-evidently consistent theory would have to be very weak indeed. Rather (10) is something that has been arrived at by following a certain heuristic train of thought, starting with the idea of limitation of size, and the additional suggestion that "too many" should be understood to mean "indefinitely or indefinably many" rather than to mean "equinumerous with all objects," as in previous attempts.

Principles of the general shape of (10), saying that whatever holds in the macrocosm of all objects holds in the microcosm of elements of some set t, are called *reflection principles*. Reflection principles were introduced into axiomatic set theory by Azriel Levy, and especially developed by Paul Bernays (1961), who showed they could be of great power. (Neither Levy nor Bernays is directly concerned with *motivating* reflection principles as axioms. Levy derives one from the axioms of ZF, and Bernays explores the consequences of taking one as an unargued assumption.) The reflection principle (10) is indeed tantalizingly close to one studied by Bernays. The Bernays principle differs from (10) only by imposing a certain extra technical condition on t.

The task of going back over the heuristic line of thought leading to (10) and trying to motivate the extra technical condition by special pleading seems daunting, and fortunately it is unnecessary, since it turns out to be possible to derive the Bernays principle from (10) using a technique of Bernays himself, combined with the one distinctively Fregean feature of our axiomatization, the subordination axioms, which have not yet been used.

The Bernays technique, applied to (10), is simply this, that if we have proved a theorem θ, then any ϕ implies θ & ϕ, and applying (10) to this conjunction rather than just ϕ, we get $\phi \rightarrow \exists t\ (\theta \ \&\ \phi)^t$, which is the same as $\phi \rightarrow \exists t\ (\theta^t \ \&\ \phi^t)$. If θ^t implies some special property $\psi(t)$, then we get $\phi \rightarrow \exists t\ (\psi(t) \ \&\ \phi^t)$. *If the relativization of a theorem to t implies a special property of t, then the reflection principle can be strengthened to require t to have this special property.* Since the conjunction of any number of theorems is still a theorem, the theorem-relativization principle can be used over and over to get more and more special properties, giving more and more theorems, giving more and more special properties, a feature in connect with which Bernays uses the expression "self-reinforcing."

Let us first apply this idea taking as our θ the subordination axiom (1). Its relativization reads as follows:

$$\forall x(x \in t \rightarrow (ßx \leftrightarrow \exists X \subseteq t\ €xX))$$

The useful part of the biconditional is the implication from left to right. For suppose $x \in t$ and $y \in x$. Then this direction of the biconditional tells us there is an $X \subseteq t$ such that $€xX$. The subordination axiom (2) tells us that since $y \in x$, we have Xy, and this together with $X \subseteq t$ tells us we have $y \in t$. Any element of an element of t is an element of t. As was mentioned in the discussion of large cardinals in chapter 1, a set with this special property is called *transitive*. The special property of transitivity is the "extra technical condition" in the Bernays reflection principle, which amounts to what we have just proved, namely:

(11) $$\phi \rightarrow \exists t(t \text{ transitive } \& \ \phi^t)$$

Let me call the set theory in the monadic second-order language with just ß and $\in$ and the axioms of extensionality, separation (in a single-axiom, second-order version), and reflection in the stronger form (11), *Bernays set theory* B. For though not precisely the system with which Bernays worked, it is nearly enough so for present purposes. And let me call the set theory that adds € and the subordination axioms, and substitutes reflection in the weaker form (10), *Fregeanized Bernays set theory* FB.

Bernays in effect proved that all the remaining existence axioms of second-order Zermelo-Frankel set theory ZF^2 as well as several large cardinal axioms (namely: pairing, union, power, infinity, replacement, inaccessibles, hyperinaccessibles, . . .) can be deduced in B. As an immediate corollary of our deduction of (11) from (10) and the subordination axioms, we have the following:

> ***Fregeanization of the Theorem of Bernays.*** All the existence axioms of second-order Zermelo-Frankel set theory as well as inaccessibles, hyperinaccessibles, . . . are deducible in Fregeanized Bernays set theory.

It has generally been my policy not to reproduce proofs already readily available in the literature, except where it is necessary to review the proof in order to obtain refinements. The only refinement I have to offer to the theorem of Bernays, apart from the initial heuristic motivation for the reflection principle, is the observation that the extra technical condition of transitivity does not need to be assumed, but can be deduced using the Fregean subordination axioms; and this refinement has already been presented. The Bernays proof is so beautiful, however, that it is difficult to refrain from presenting at least part of it, and to do so may facilitate understanding of what has gone before.

We got a refinement of reflection using the subordination axioms and extensionality, but have not yet tried reflecting the separation axiom. Here "separation" may be any convenient single-axiom formulation of the principle, since all the various formulations are theorems. The most convenient to consider gives the following as its relativization:

(12) $\forall x \in t\, \forall X \subseteq t\, \exists y \in t\, \forall z \in t\, (z \in y \leftrightarrow z \in x\ \&\ Xz)$

Here t may be taken to be transitive. Now if $x \in t$ and $w \subseteq x$, first note that $x \subseteq t$ by transitivity. So we have $w \subseteq t$, and therefore the concept $X =$ «z: $z \in w$» satisfies $X \subseteq t$. So relativized separation (12) tells us there is a $y \in t$ for which we have the following:

(13) $\forall z \in t\, (z \in y \leftrightarrow z \in x\ \&\ Xz)$

The relativization in the quantifier $\forall z \in t$ in (13) is redundant, since the condition $z \in y$ on the left side and the condition $z \in x$ on the right side of the biconditional in (13) both imply $z \in t$ already, given that $x \in t$ and $y \in t$ and t is transitive. So what we *really* have is the following:

(14) $\forall z(z \in y \leftrightarrow z \in x\ \&\ Xz)$

But recalling that $X =$ «z: $z \in w$» and $w \subseteq x$, the conjunction $z \in x\ \&\ Xz$ on the right side of the biconditional in (14) is equivalent to $z \in w$, and so we have $w = y \in t$. In other words, every subset w of an element x of t is an element of t. As was said in the discussion of large cardinals in chapter 1, a transitive set with this special property is called *supertransitive*. So we have established the following super-Bernays version of reflection:

(15) $\phi \rightarrow \exists\, t(t \text{ supertransitive } \&\ \phi^t)$

To derive pairing, we only need the original version of reflection. Given u and v, we reflect the following logical truth:

(16) $\exists x(x = u)\ \&\ \exists\, y(y = v)$

to get a set t such that (16) holds relativized to t, which is to say, the following holds:

(17) $$\exists x \in t\,(x = u)\ \&\ \exists y \in t\,(y = v)$$

But (17) is just a long-winded way of saying $u \in t$ and $v \in t$. We may now apply separation to separate out from t just those of its elements that are either identical with u or identical with v, and so obtain the pair $\{u, v\}$.

For union, we use the Bernays version with transitivity to show that for every u there is a transitive set to which it belongs. Such a t will contain all the elements of u and all the elements of those elements, and we can separate out the elements of elements to get the union $\bigcup u$.

For power, we use the version with supertransitivity to show that for every u there is a supertransitive and transitive set t with $u \in t$. Such a t will contain all the subsets of u, and we can separate out those subsets to get the power set $\wp(u)$.

For infinity and replacement I refer the reader to the work of Bernays himself. For inaccessibles, let me just say that one of the many known equivalent formulations of the axiom is the existence of a supertransitive and transitive set t such that all the existence axioms (with separation and replacement in their second-order, single-axiom forms) hold relativized to t. But this we get immediately by reflection as soon as we have got the existence axioms themselves.[13]

THERE REMAIN ONE AXIOM of ZFC that we have not got in its usual form, and two that we have not got at all, and something must be said about these and how to get them. The one we don't have in its usual form is extensionality, the other two are foundation or regularity, and choice.

As for extensionality, the form in which we have it and the usual form are as follows:

(18) $$ßx\ \&\ ßy\ \&\ \forall z(z \in x \leftrightarrow z \in y) \rightarrow x = y$$

(19) $$\forall z(z \in x \leftrightarrow z \in y) \rightarrow x = y$$

Our form (18) is the only reasonable form if one assumes, as Frege himself always did, that the object variables range over absolutely all objects. The usual adoption of form (19) is not an expression of the unreasonable belief that absolutely every object is a set, but rather of an intention to impose an un-Fregean restriction on the range of the object variables. They are to be restricted to sets, and not just to sets, but to sets of which the elements are all of them sets, and are not just sets, but are sets whose elements are all of them sets, and are not just. . . . In short, the restriction is to sets x such that every object standing in the ancestral of the elementhood relation to x is a set. Such are called *pure* sets.

The obvious strategy for getting the usual version (19) of extensionality would be to relativize quantifiers to pure sets. To do this we must first find a formula $\theta(x)$ expressing that x is pure. This is no problem using second-order logic, where we have Frege's definition of the ancestral. But in fact there is an equivalent first-order $\theta(x)$, which simply says that there is a transitive t such that $x \in t$ and every element of t is a set. Since a transitive t such that $x \in t$ will have all objects standing in the ancestral of the elementhood relation to x as elements, if all elements of t are sets, x is certainly pure. Conversely, we know that every x is an element of some transitive set t and we can separate out from t just x and the objects that stand in the ancestral of the elementhood relation to x to form a set u that will be transitive and will have $x \in u$, and that if x is pure will have only sets as elements.

We need to check that in getting extensionality in version (19) we have not lost anything else. We need to check, for instance, that the axiom of pairing still holds when quantifiers are relativized to pure sets. If there exist transitive t and s containing u and v, respectively, and both having only sets as elements, we need to show there is a transitive r containing $w = \{u, v\}$ and having only sets as elements. This can be done by verifying that $r = s \cup t \cup \{w\}$ will do, but then the same has to be done for every other set-existence axiom. Moreover, to get the axiom of foundation, one needs to do something similar, with a second round of verifications. Fortunately the details are given in introductory textbooks and need not concern us here.[14]

As for choice, which has many equivalents, such as the principle that every set can be well-ordered, there are several options. The first option is to obtain choice by relativization to Gödel's constructible sets—this is how Gödel showed ZFC can be interpreted in ZF. The procedure is artificial, since in contrast to the case of extensionality, those adopting AC are *not* expressing an intention to limit the range of the object variables to constructible sets. But artificiality is not always an objection. One should be aware, though, that if one is prepared to get axioms by relativization, then an ostensibly much more "minimalistic" starting point than FB will get us to the same goal. This has been established in recent unpublished work of Friedman, which I will digress to describe briefly. Friedman first considers a language with only object variables and only $\in$, no second-order apparatus at all, and not even $=$. He considers a first theory I will call HFR_1, for *Harvey Friedman reflective set theory, first version*, having no axioms except the following scheme:

$$\phi^\theta \rightarrow \exists t(\forall x(x \in t \rightarrow \theta(x)) \ \& \ \phi^t)$$

If ϕ holds relativized to θ, then there is a t such that θ holds for all elements of t and ϕ holds relativized to t. He considers also three successively stronger theories I will call HFR_i for $i = 2, 3, 4$, each again having a single axiom scheme in the form of a reflection principle. He shows that HFR_1 is equiconsistent with analysis P^2 or ZF^-, HFR_2 with ZFC, HFR_3 with B or FB, and HFR_4 with a still stronger theory having still larger large cardinals. The proofs are technical *tours de force*, and beyond the scope of this monograph.

A second option is, now that we have pairing, to (re)introduce dyadic second-order logic, coding two-place relational concepts by one-place concepts with ordered pairs falling under them, and then adopt von Neumann's axiom, which can be motivated as another expression of the same underlying principle of limitation of size that motivates separation and reflection. Choice then follows by a clever argument of von Neumann. (He develops a theory of ordinals in the way that is now standard in axiomatic set theory; he uses the Burali-Forti paradox to conclude that the ordinals do not form a set; he uses his axiom to conclude that there is a one-to-one correspondence between all objects and the ordinals; and lastly, he uses the one-to-one correspondence and the fact that the ordinals are well-ordered to establish that any set can be well-ordered.)

A third option is to do what Bernays himself did, and adopt Hilbert's ε-symbol, which in effect builds AC into *first*-order logic. To describe this option in detail would take us too far afield.

A fourth option is to do what is done in most conventional axiomatizations, and simply adopt AC as an axiom that has to be independently motivated.

3.7 Second-Order Logic Reconsidered

Rather than elaborate further on the set-theoretic material I have alluded to, most of which is readily available elsewhere, and very little of which has much savoring of Frege about it, I will use the remaining space available to me to cast a more critical eye upon the tool employed in one way or another in all recent Frege-inspired projects, second-order logic itself.

With first-order logic we have a syntactic or proof-theoretic notion of a conclusion, perhaps a conjectured theorem, being *deducible* from a premiss, perhaps a conjunction of axioms of some theory. Special or degenerate cases give us the notions of *inconsistency* and *demonstrability*: a premiss is inconsistent if and only if any conclusion is deducible from it, and a conclusion is demonstrable if it is deducible from any premiss. If we start with the notion of inconsistency, we can alternatively define deducibility of ψ from ϕ as the inconsistency of $\sim\psi$ & ϕ, and demonstrability of ψ as the inconsistency of $\sim\psi$.

We also have a semantic or model-theoretic notion of a conclusion being a *consequence* of a premiss: ψ is a consequence of ϕ if and only if every model that makes ϕ true makes ψ true. We also have related notions of *unsatisfiability* and *validity*: a premiss is unsatisfiable if true in no model, and a conclusion is valid if true in all models. If we start with the notion of unsatisfiability, we can alternatively define ψ being a consequence of ϕ as the unsatisfiability of $\sim\psi$ & ϕ, and validity of ψ as the unsatisfiability of $\sim\psi$.

The two triples of notions coincide by the (soundness theorem and the) Gödel completeness theorem: θ is satisfiable if (and only if) θ is consistent. From this it follows also that deducibility coincides with consequence, and demonstrability with validity. The connection between the syntactic and semantic notions means that model-theoretic results can be used to derive proof-theoretic corollaries, and vice versa.

Thus far I have considered *only* syntactic notions in connection with second-order theories, and have emphasized that so far as consistency, deducibility, and demonstrability are concerned, second-order logic is just many-sorted first-order logic, and all the tools used in connection with the latter can be used for the former. This includes the use of models, in the sense of *first*-order logic, to establish consistency. (Sometimes there are advantages to establishing consistency proof-theoretically rather than model-theoretically, because a proof-theoretic argument for consistency typically gives additional information beyond mere consistency in a way that a model-theoretic argument typically does not. But a model-theoretic proof of consistency does prove consistency, and a theory doesn't become any more consistent than it already was when a model-theoretic proof is replaced by a proof-theoretic proof.)

In working with models for second-order logic, it is always sufficient to consider only ∈-models, or models where the domain of the second-order variables consists of subsets of the domain of the first-order variables (since it is an easy theorem that an arbitrary model is isomorphic to an ∈-model, so long as one is assuming extensionality, as I always have been). Sometimes the domain of the second-order variables consists of *all* subsets of the domain of the first-order variables; often it does not. In the former case we have a *standard* or *Tarski* model, in the latter case a *general* or *Henkin* model. General models are just as good as standard models for establishing facts about syntactic consistency, deducibility, or demonstrability.

There is also, however, a notion of standard semantic consequence, of which I have said nothing so far, according to which ψ is a *standard* consequence of ϕ if and only if ψ is true in every *standard* model in which ϕ is true. With this notion come associated notions of *standard* unsatisfiability or satisfiability, and *standard* validity or invalidity. And in (agreement with the soundness theorem and in) contrast to the Gödel completeness theorem for first-order logic, it is known that standard consequence is (a necessary but) not a sufficient condition for deducibility, and similarly for standard unsatisfiability and inconsistency and standard validity and demonstrability. This is because formula θ may have *only* general models and no standard models in which it is true.

This is so for *any* system of rules of deducibility, in whatever format, so long as it is required that the rules be recursive (which by Church's thesis is a necessary condition for its being effectively decidable whether one has followed the rules or not in any given case, which must be so for any rules worthy of the name). For then the set of inconsistent formulas will be recursively enumerable, while the set of standardly unsatisfiable formulas is known not to be, and the mismatch between (in)consistency and standard (un)satisfiability gives rise to a similar mismatch between (un)demonstrability and standard (in)validity and between deducibility and standard consequence. Such is the notorious *incompleteness* (or rather, since it applies to *any* system of rules of deducibility, *incompletability*) of second-order logic.

Three questions are raised by this result, which is a kind of corollary to the Gödel incompleteness theorems, though the route from the one to the other need not concern us here. First, what does *standard* satisfiability have to do with *intuitive* satisfiability, which is to say, with satisfiability in the intuitive sense of coming out true under some interpretation of what objects the object variables are supposed to range over, and what relations the relation symbols of the language are supposed to denote? Second, if standard satisfiability does coincide with intuitive satisfiability, so that the incompleteness of second-order logic means that intuitive satisfiability does *not* coincide with syntactic deducibility, is this a cogent ground for objecting to the use of second-order logic? Third, if not, is second-order logic, especially as conceived by Frege and more recent writers influenced by him, open to any other cogent objection? In this section I will take up the three questions in the order listed.

IN GENERAL, WHEN DOING MATHEMATICS of any kind, including meta-mathematics (proof theory and model theory), one is informally working in some kind of background mathematical meta-theory. And when doing model theory, generally this meta-theory is some kind of set theory (though when doing only proof theory, it may be something much weaker, even as weak as $Q_4 = I\Delta_0$(superexp) or weaker). The default assumption when no meta-theory is explicitly indicated is that one is conforming to generally accepted standards in contemporary mathematics, which is to say that one is working within the generally accepted set theory of contemporary mathematics, ZFC (though often one is really using far less than the full strength thereof). It turns out that it makes a real difference to the question of the relationship between standard satisfiability (and consequence and validity) and intuitive satisfiability (and consequence and validity) just what set theory one is working in, *and the set theory* FB *is exactly what is needed to be able to give a decisive answer to the question of the relation of standard to intuitive validity*. This fact should certainly be a point in favor of that set theory for those committed to working with second-order logic. This point has in effect been noted by Boolos (1985a) in a slightly different context, as well as in Shapiro (1987), which contains a general discussion of second-order reflection principles, and has been examined in depth by Rayo and Uzquiano (1999), but the point may still not be as widely known and understood as it should be, and deserves reiteration and elaboration.

For simplicity, let me consider just a *monadic* second-order language, with a single non-logical symbol, say a two-place relation symbol ® applying to object variables. Let me review in a little more detail what is meant by having a standard model in this context, before comparing that notion of standard satisfiability with the intuitive notion.

We begin with Tarski's notion of a model for the first-order part of the language. Such a model consists of a non-empty set $|\mathfrak{M}|$, the *universe* of the model, the range of the object variables, and a set $®^{\mathfrak{M}}$ of ordered pairs of elements of $|\mathfrak{M}|$, the *relation* of the model. In order to define what it is for a closed formula with no free variables to be true in the model, one must define what it is for a formula *with or without* free variables to be satisfied or true *relative to an assignment* of elements of the universe of the model to any free variables. The details of the definition, which proceeds by induction on the complexity of the formula, will not be recalled, except to mention that at the base step, ®*uv* is true for the assignment of *m* and

n to u and v if and only if the ordered pair $\langle m, n\rangle$ belongs to the set $®^{\mathfrak{M}}$, while at the induction step for the universal quantifier, $\forall u\psi(u)$ is true if and only if $\psi(u)$ is true whatever element m of $|\mathfrak{M}|$ is assigned to u. In (one version of) the usual symbolism:

(1) $\mathfrak{M}, m, n \models ®uv$ iff $\langle m, n\rangle \in ®^{\mathfrak{M}}$

(2) $\mathfrak{M} \models \forall u\psi(u)$ iff $\mathfrak{M}, m \models \psi(u)$ for all $m \in |\mathfrak{M}|$

For monadic second-order logic the Tarski definition of standard model is exactly the same, a set with a set of ordered pairs of its elements. Second-order variables are taken to range over subsets of $|\mathfrak{M}|$, and new clauses have to be added to the definition of truth, for the new atomic formulas of the form Uu and for the second-order quantifiers. They read as follows:

(3) $\mathfrak{M}, M, m \models Uu$ iff $m \in M$

(4) $\mathfrak{M} \models \forall U\psi(U)$ iff $\mathfrak{M}, M \models \psi(U)$ for all $M \subseteq |\mathfrak{M}|$

Standard satisfiability amounts to truth in some standard model.

Now consider a formula $\phi(®)$, say one beginning as follows:

(5) $$\forall U\exists v(Uv \;\&\; ®vv \;\&\; \ldots)$$

What it means intuitively to say that (5) is satisfiable is that there are some objects and some relation on them such that (5) comes out true if the object variables u are taken to range over all and only those objects, the concept variables U are taken to range over all and only those concepts under which such objects fall, and the symbol ® is taken to stand for the relation in question. That is to say, (5) is intuitively satisfiable if and only if the following formula is true:[15]

(6) $$\exists X \exists R \forall U \subseteq X \exists v(Xv \;\&\; Uv \;\&\; Rvv \;\&\; \cdots)$$

The standard definition differs from this intuitive definition in two respects. First, it insists that the range of the first-order variables must be a *set*; and second, it takes the concept variables to range over sub*sets* rather than sub*concepts* of that set. Thus (5) is standardly satisfiable if and only if the following is true:

(7) $\exists x\, \exists r\, \forall u \subseteq x\, \exists v(v \in x \,\&\, v \in u \,\&\, \langle v, v\rangle \in r \,\&\, \cdots)$

The second of the two points of difference noted between the intuitive and the standard notion is not very important. If the range of the first-order variables is a set x, talking about sub*sets* thereof is equivalent to talking about sub*concepts* thereof. For on the one hand, every $u \subseteq x$ determines a $U \subseteq x$, namely «v: $v \in u$», and conversely every $U \subseteq x$ determines a $u \subseteq x$, namely $\{v \in x\text{: } Uv\}$. This last point does, however, assume the axiom of separation is available in the background set theory in which we are conducting the current discussion. But assuming separation, talk of subsets and talk of subconcepts are equivalent, and similar remark applies to relations, so (7) is equivalent to the following:

(8) $\exists\, x\, \exists\, R\, \forall\, U \subseteq x\, \exists\, v \in x\, (Ux \,\&\, Rvv \,\&\, \cdots)$

The first of the two points of difference noted between the intuitive and the standard notion is the important one. The intuitive requirement is that (6) should be true, while the standard requirement is that (8) should be true, which a little thought shows amounts to the requirement that (6) should be true *relativized to some set x*. Now the latter requirement certainly implies the former, since if we have the set x of (8) we can simply take «v: $v \in x$» as the X in (6). It is the converse implication that is problematic. But *if* we assume the axiom of reflection is available in the background set theory in which we are conducting the current discussion, *then* these two requirements are the same, and the intuitive and standard notions coincide after all. For if θ is the formula displayed in (6), reflection tells us precisely that $\theta \rightarrow \exists x\ \theta^x$.

This is what I meant by saying earlier that the set theory FB is exactly what is needed to be able to give a decisive answer to the first of our three questions about the status of second-order logic, that of the relation between standard satisfiability and intuitive satisfiability. The two set-existence axioms of FB, separation and reflection, are exactly what is needed to get the equivalence of (6)–(8).

LET US TURN NOW to the second of the three questions about the status of second-order logic with which we began. Is it a point against second-order logic that standard (un)satisfiability and syntactic (in)consistency, and hence standard consequence and syntactic deducibility, and standard (in)validity and syntactic (un)demonstrability diverge?

It might be thought to be so for the following reason. We might come to have what we regard as good reason for thinking some formula of second-order logic is intuitively valid, even though it is not deducible from our logical axioms of comprehension (and extensionality) by the ordinary rules of deduction for second-order logic (which are just those of many-sorted first-order logic). In that case, presumably we would want to add the formula as a new logical axiom that should have been assumed from the beginning had we been aware of it at the time.

To give concreteness to this abstract possibility, consider the logical or so-called *global* form of the axiom of choice, which using the equivalent well-ordering version may be expressed as follows:[16]

(10) $$\forall X \exists R (R \text{ well-orders } X)$$

The global axiom of choice may well be something we convince ourselves is intuitively valid (perhaps by something like an informal counterpart of von Neumann's argument). And if we do so, it would seem that we ought to add global choice as a new logical axiom, in addition to comprehension (and extensionality). That would amount to changing the rules of deduction in the middle of our ongoing theorem-proving activity, which may seem objectionable in the way that changing the rules of game in the middle of play is objectionable.

Now we know from the Gödel incompleteness theorems that, even keeping logic fixed, and even working only with first-order logic, as when working in a set theory like ZF, we might come to have good reason to think some formula of the language of set theory is intuitively true, even though it is not deducible from the set-theoretic axioms of ZF. The Gödel theorems suggest the set-theoretic statement coding the statement that ZF is consistent as an example, but historically an example was the set-theoretic or *local* version of the axiom of choice, which in its equivalent well-ordering version reads as follows:

(11) $$\exists x \exists r (r \text{ well-orders } x)$$

The local axiom of choice is something that set theorists initially distrusted, then convinced themselves is intuitively true (some of them by an informal counterpart of von Neumann's argument). And when they did so, they added, as it would seem they ought to have done, local choice AC as a new set-theoretic axiom, in addition to the axioms of ZF, so obtaining ZFC. Does not *this* amount to changing the rules about what premisses can be taken for granted in a deduction in the middle of their ongoing theorem-proving activity? And why would adopting global choice be more objectionable than that?

Nonetheless, there is in some of us a stubborn feeling that changing the logical axioms is more undesirable than changing the set-theoretic axioms, which since Gödel we have come to accept as something inevitable—and even something desirable, for those who think set theorists are in the process of gaining a deeper and deeper intuitive understanding that may motivate stronger and stronger axioms.

To those who share this stubborn feeling, the news may be welcome that *the set theory* FB *is* exactly *what is needed to make any future adoption of new* logical *axioms unnecessary*, even if we become convinced that certain formulas not previously accepted as axioms are intuitively true. For the reflection principle implies that adopting something like (10) as a logical axiom is equivalent to adopting something like (11) as a set-theoretic axiom. Any violation of global choice would by reflection yield a violation of local choice. This is just a particular case of what we saw in the preceding section, that any question of intuitive validity reduces to a question of standard validity, which is a question about the truth of a certain first-order set-theoretic formula.

Thus once one has accepted the set-theoretic axioms of FB, there will never be occasion to assume new logical as opposed to set-theoretic axioms, no matter what more we become convinced of, because adopting a new logical axiom would always be equivalent to adopting a corresponding new set-theoretic axiom, and we can do the latter rather than the former. In a sense the set-theoretic axioms of FB imply that the rules of deduction for second-order logic that we already have are *as complete as they need to be.*[17]

To readers who have previous acquaintance with the controversies over the status of second-order logic, I suppose I should offer apologies for not addressing well-known objections sooner. But the reason for my delaying doing so until after the introduction of the set theory FB should now be plain.

THERE REMAINS AN OBJECTION directed specifically against a Fregean understanding of second-order logic that I have delayed raising until the wide range of modified Fregean systems was before us, enabling us to judge the effect of the objection on various ones of them. The objection is one that many semi-Fregeans, quasi-Fregeans, neo-Fregeans, and the like consider it bad form to mention, but mention it I must, because unlike prefix-Fregeans I am convinced that it is cogent, and has a devastating effect on many of the theories I have surveyed. The objection is a simple grammatical one that has been repeatedly urged, in slightly different form, by Quine, and repeatedly ignored.

The background is as follows. It is crucial to second-order logic that the items over which the second-order variables range are not in the range of the first-order variables. The unrestricted assumption of extensions, which does not even quite put them there, but just puts representatives of them there, leads to Russell's paradox, after all. Now for Frege, *anything that can be denoted by a singular noun phrase* is an object and therefore in the range of the first-order variables. So for Frege it is crucial that concepts, which are the denotations of predicates or verb phrases requiring a singular noun phrase to make a complete sentence, cannot be denoted by noun phrases. Thus the concept denoted by "is a horse" cannot be denoted by any noun phrase—*not even this one:*

the concept denoted by "is a horse"

This is an embarrassment. Frege asked his readers not to begrudge him a grain of salt in this connection, and many commentators regard it as uncharitable and ill-bred to allude to this problem. I have, of course, just now done more than merely allude to it. My excuse for this display of lack of charity and good breeding is that I believe Frege's embarrassment over the concept of being a horse, though perhaps trifling in itself, is a symptom of a much more serious problem. And to ignore this more serious problem would be to allow Fregeans not just to have a grain of salt, but to steal a whole salt mine.

The problem is that there is no explanation of the meaning of $\forall X$ and $\exists X$ intelligible to one who speaks only English (or German), and not *Begriffsschrift* (if one may speak of "speaking" a non-linear notation), that is compatible with regarding X as standing strictly for a verb phrase or predicate and not a noun phrase. Generally one is told that $\forall$ and $\exists$ may be read as "for all . . . " and "for some . . . " or some other such ordinary locution. But these ordinary-language locutions *all demand to be followed by noun phrases*.

The problem becomes apparent if one tries to pronounce a second-order formula such as the law of indiscernibility of identicals:

(12) $$x = y \rightarrow \forall X(Xx \leftrightarrow Xy)$$

According as one has or has not been raised as a Fregean, one is likely to come out with one or the other of the following:

(12a) If x is the same thing as y, then x falls under every concept that y does, and vice versa.

(12b) If x is the same thing as y,
then x has every property that y does, and vice versa.

Nouns "concept" or "property" are spontaneously inserted in place of an X that is supposed to be standing for a verb phrase. The reason this happens is that the sort of thing one would have to do to stick to verb phrases produces a result that is not intelligible or even grammatical English:

(12c) *If x is the same thing as y,
then everyso, x is so if y is so, and vice versa.

To me, at least, this simple grammatical problem seems to make Fregean second-order logic as much nonsense as the non-word "everyso."

A partial solution to the problem has been provided by Boolos (1984), who in effect observes that while "for all . . . " and "for some . . . " may demand nouns, they don't demand *singular* ones. He in effect proposes that singular quantifications over concepts, "there exists a concept X such that. . . ," can be replaced by plural quantifications over objects, "there are some objects, the xs, such that. . . ." How would the various modified Fregean systems I have surveyed, and that are set out in table J, fare if we abandoned Frege's second-order logic for Boolos's plural first-order logic?

For the most part, not well. For one thing, if we follow Boolos in regarding the ordinary plural locutions as intuitively intelligible prior to theories of sets, classes, concepts, or anything of the kind, there seems to be no way to motivate predicativist restrictions on comprehension, and theories involving such restrictions, such as were surveyed in the preceding chapter. For plural comprehension just says that, for any formula $\phi(u)$, there are some objects, the xs, such that any object u is among them if and only if $\phi(u)$. And *of course* there are such objects, namely, the objects for which ϕ holds.

For another thing, plural logic only provides a replacement for *monadic* second-order logic, and leaves us without a way to formulate most of the theories considered in this chapter, namely, all those to which the notion of one-to-one correspondence (not explained in terms of a prior notion of ordered pair) was crucial.[18]

This effectively leaves only FB, the most important and almost the only theory based on impredicative but monadic second-order logic, to be considered in our survey. Of course, by the time one adopts any version of the limitation of size idea, one has adopted the central idea of Frege's rival Cantor, an idea of which there is no trace in Frege himself. And by the time one adopts the specific version of the limitation of size idea embodied in FB, there is very little residue of Frege left. Nonetheless, the Fregean residuum was crucial to our ability to get by with a less technical and easier-to-motivate version of reflection, and we must consider whether the residual Fregean idea involved survives translation into plural logic.

The Fregean idea was that the relationship of element *z* to set *x* is subordinate to or derivative from more primary relationships that a mediating concept *Y* bears to the two objects *z* and *x* separately, namely, the relationships of *z* falling under *Y* and of *x* being the extension of *Y*. The plural translation would be the idea that the relationship of element *z* to set *x* is subordinate to or derivative from more primary relationships between some mediating objects, the *ys*, to the two objects *z* and *x* separately, namely, the relationships of *z* being one of the *ys*, and of *x* being the set of the *ys*. In other words, the relationship of a set to the things of which it is the set, taken collectively, has priority over the relationship of the set to any particular one of those things, taken individually. The issue is subtle, even elusive, but the translated Fregean idea does not appear unreasonable—or un-Cantorian. It would be possible to translate the whole above discussion of Fregeanized Bernays set theory FB into a discussion of an analogous system, Boolosified Bernays set theory BB, where no Fregean idea, but only the ghost of one, would remain. But a monograph on recent modified Fregean systems is not the place for a discussion of such a system.[19]

Tables

TABLE A Fregean Categories in Ajdukiewicz/Bar-Hillel Notation

Type Symbol	Example	Referent	Variables
S	Socrates is wise	*truth-value*	
N	Socrates	*object*	*x*, *y*, …
S/N	. . . is wise	*concept*	*X*, *Y*, …
S/NN	. . . taught __	*relational concept*	*R*, *S*, …
S/(S/N)	Someone . . .	*higher concept*	**X**, **Y**, …
S/(S/N)(S/N)	Whoever . . . , __	*higher relational concept*	**R**, **S**, …

TABLE B Axiomatizations of Frege's Set Theory

Extension-of Relation Symbol € Primitive	
Existence Assumption	$\exists x$ €xX
"Law V"	€xX & €$yY \rightarrow (x = y \leftrightarrow X \equiv Y)$
Definition ‡/€	‡$X = \iota x$€xX
Definition ß/€	ß$y \leftrightarrow \exists Y$ €yY
Definition ∈/€	$x \in y \leftrightarrow \exists Y$ (€yY & Yx)
Extension-of Function Symbol ‡ Primitive	
Existence Assumption	Hidden
"Law V"	‡X = ‡$Y \leftrightarrow X \equiv Y$
Definition €/‡	€$xX \leftrightarrow x$ = ‡X
Definition ß/‡	ß$y \leftrightarrow \exists Y$ (y = ‡Y)
Definition ∈/‡	$x \in y \leftrightarrow \exists Y$ (y = ‡Y & Yx)
Definition { : }/‡	$\{x$: $\phi(x)\}$ = ‡«x: $\phi(x)$» = ‡$\iota Y \forall x(Yx \leftrightarrow \phi(x))$
Sethood/Elementhood Relation Symbols ß, ∈ Primitive	
Existence Assumption	$\exists y$(ßy & $\forall x(x \in y \leftrightarrow \phi\ (x)))$
"Law V"	$\forall y \forall z$(ßy & ßz & $\forall x(x \in y \leftrightarrow x \in z) \rightarrow y = z)$
Definition €/ß∈	€ $yX \leftrightarrow$ ßy & $\forall x(x \in y \leftrightarrow Xx)$
Definition ‡/ß∈	‡$X = \iota y$(ßy & $\forall x(x \in y \leftrightarrow Xx))$
Definition { : }/ß∈	$\{x$: $\phi(x)\} = \iota y$(ßy & $\forall x(x \in y \leftrightarrow \phi(x)))$

Set-of Variable-Binding Term-Forming Operator { : } Primitive

Existence Assumption	Hidden
"Law V"	$\{x: \phi(x)\} = \{x: \psi(x)\} \leftrightarrow \forall x(\phi(x) \leftrightarrow \psi(x))$
Definition €/{ : }	€$xX \leftrightarrow x = \{w: Xw\}$
Definition ‡/{ : }	‡$X = \{w: Xw\}$
Definition ß/{ : }	ß$y \leftrightarrow \exists Y(y = \{x: Yx\})$ ∴ Comprehension $\phi \rightarrow$ ß$\{x: \phi(x)\}$
Definition ∈/{ : }	$w \in y \leftrightarrow \exists Y(y = \{x: Yx\}$ & $Yw)$ ∴ Comprehension $\phi \rightarrow (w \in \{x: \phi(x)\} \leftrightarrow \phi(w))$

For verbal descriptions, see §1.2.

TABLE C Basic Notions and Notations of Set Theory

TERMINOLOGY	SYMBOL	DEFINITION
Null Set	$\varnothing$	$= \{x: x \neq x\}$
Singleton	$\{u\}$	$= \{x: x = u\}$
(Unordered) Pair	$\{u, v\}$	$= \{x: x = u \vee x = v\}$
(Unordered) Triple	$\{u, v, w\}$	$= \{x: x = u \vee x = v \vee x = w\}$
Adjunction	$u \wedge y$	$= \{x: x \in u \vee x = y\}$
Deletion	$u \setminus y$	$= \{x: x \in u \;\&\; x \neq y\}$
(Simple) Union	$u \cup v$	$= \{x: x \in u \vee x \in v\}$
(Simple) Intersection	$u \cap v$	$= \{x: x \in u \;\&\; x \in v\}$
(Grand) Union	$\bigcup u$	$= \{x: \exists y(y \in u \;\&\; x \in y)\}$
(Grand) Intersection	$\bigcap u$	$= \{x: \forall y(y \in u \rightarrow x \in y)\}$
Difference	$u - v$	$= \{x: x \in u \;\&\; x \notin v\}$
Subset	$u \subseteq v$	$\leftrightarrow \forall x(x \in u \rightarrow x \in v)$
Power Set	$\wp(u)$	$= \{y: y \subseteq u\}$
Complement	$-y$	$= \{x: x \notin y\}$
Universal Set	V	$= \{x: x = x\}$

TABLE D Axioms of ZFC

NAME	FORMULA
Extensionality	$\forall z(z \in x \leftrightarrow z \in y) \rightarrow x = y$
Null Set	$\exists x \forall y(y \in x \leftrightarrow y \neq y)$
Pairing	$\exists x \forall y(y \in x \leftrightarrow y = u \vee y = v)$
Union	$\exists x \forall y(y \in x \leftrightarrow \exists z(z \in u \ \& \ y \in z))$
Power Set	$\exists x \forall y(y \in x \leftrightarrow \forall z(z \in y \rightarrow z \in u))$
Infinity	$\exists x(\varnothing \in x \ \& \ \forall y(y \in x \rightarrow y' \in x))$
Separation	$\exists x \forall y(y \in x \leftrightarrow y \in u \ \& \ \phi(y))$
Foundation	$\exists x(x \in u) \rightarrow \exists x(x \in u \ \& \sim \exists y(y \in x \ \& \ y \in u))$
Replacement	$\forall v \exists ! w \psi(v, w) \rightarrow \exists x \forall y(y \in x \leftrightarrow \exists v(v \in u \ \& \ \psi(v, y))$
Choice	$\forall u((\forall x(x \in u \rightarrow \exists u(u \in x) \ \& \ \forall x \forall y(x \in u \ \& \ y \in u \ \& \sim x = y \rightarrow \sim \exists w(w \in x \ \& \ w \in y)) \rightarrow \exists z \forall x(x \in u \rightarrow \exists ! w(w \in x \ \& \ w \in z))$

For verbal descriptions, see §1.5d.

TABLE E Twenty Milestones on the Fundamental Series

1ST-ORDER ARITHMETICS	2ND-ORDER ARITHMETICS	1ST-ORDER SET THEORIES
$Q_2 = I\Delta_0$		
$Q_3 = I\Delta_0(\exp)$		
$Q_4 = I\Delta_0(\sup)$		
$Q_\omega \approx$	$F_0 = RCA_0 \approx$	
$I\Sigma_1$	$F_1 = WKL_0$	
$I\Sigma_2$		
$P^1 = PA^1$	$F_2 = ACA_0$	$Z^- - \infty$
	$F_3 = ATR_0$	
	$F_4 =$	
	$\Pi^1_1\text{-}CA_0$	
	$\Pi^1_2\text{-}CA_0$	Z_{1^-}
	$\Pi^1_3\text{-}CA_0$	Z_{2^-}
	$P^2 = PA^2$	$Z^- \approx ZF^-$

Higher-order Arithmetics	1st-order Set Theories	2nd-order Set Theories
$P^3 = PA^3$	$ZF^- + \wp(\omega)$	
$P^4 = PA^4$	$ZF^- + \wp^2(\omega)$	
$P^\omega = PM$		
	$Z \approx ZL$	
	$ZF \approx ZFL$	NBG
		$ZFC^2 = MK$
	ZFC + inaccessibles	
	ZFC + indescribables	B
	ZFC + Reinhardt	

Note: "sup" abbreviates "superexp", "L" abbreviates "+ V =L".

For verbal descriptions of first-order arithmetics, see §1.5b; for second- and higher-order, §1.5c; for the weaker set theories, §1.5d; for the stronger, §1.5e.

TABLE F Theorems of Predicative Set Theory

NAME	FORMULA
Extensionality	ßx & ßy & $\forall z(z \in x \leftrightarrow z \in y) \rightarrow x = y$
Null Set	$\exists x($ßx & $\forall y(y \notin x))$
Singleton	$\exists x($ßx & $\forall y(y \in x \leftrightarrow y = u))$
Pair	$\exists x($ßx & $\forall y(y \in x \leftrightarrow y = u \vee y = v))$
Triple	$\exists x($ßx & $\forall y(y \in x \leftrightarrow y = u \vee y = v \vee y = w))$
Adjunction	ßu $\rightarrow \exists x($ßx & $\forall y(y \in x \leftrightarrow y \in u \vee y = v)))$
Deletion	ßu $\rightarrow \exists x($ßx & $\forall y(y \in x \leftrightarrow y \in u$ & $y \neq v)))$
Union	ßx & ßy $\rightarrow \exists z($ßz & $\forall w(w \in z \leftrightarrow w \in x \vee w \in y))$
Intersection	ßx & ßy $\rightarrow \exists z($ßz & $\forall w(w \in z \leftrightarrow w \in x$ & $w \in y))$
Difference	ßx & ßy $\rightarrow \exists z($ßz & $\forall w(w \in z \leftrightarrow w \in x$ & $w \notin y))$
Complement	ßx $\rightarrow \exists y \forall z(z \in y \leftrightarrow z \notin x)$
Universal Set	$\exists x($ßx & $\forall y(y \in x))$

TABLE G Weak Systems of Set Theory

Name	Axioms (besides Extensionality)	Interprets
PF	Null Set, Singleton, Union, Complement	$Q \therefore Q_2$
UPF	Null Set, Singleton, Pair, Triple, . . . , Complement	
ST	Null Set, Adjunction	$Q \therefore Q_2$
UST	Null Set, Pair	
UUST	Null Set, Singleton	(Q1), (Q2)
PUST	Monadic Predicative 2nd-Order Logic + UST	$Q \therefore Q_2$
P_2UUST	Dyadic Predicative 2nd-Order Logic + UUST	$Q \therefore Q_2$
PST	Monadic Predicative 2nd-Order Logic + ST	Q_3
STZ	Null Set, Adjunction, Separation	P^1
UST^2	(Full) Monadic 2nd-Order Logic + ST	P^2
$UUST^2$	(Full) Dyadic 2nd-Order Logic + ST	P^2

For verbal descriptions, see §2.1.

TABLE H Basic Notions of Predicative Second-Order Logic

Terminology	Symbol	Definition
Empty Concept	$\emptyset$	$= «x: x \neq x»$
Adjunction	$X \wedge w$	$= «x: Xx \vee x = w»$
Deletion	$X \setminus w$	$= «x: Xx \,\&\, x \neq w»$
Part	$X \subseteq Y$	$\leftrightarrow \forall x(Xx \rightarrow Yx)$
Proper Part	$X \subset Y$	$\leftrightarrow X \subseteq Y \,\&\, \exists x(\sim Xx \,\&\, Yx)$
Universal Concept	U	$= «x: x = x»$

TABLE I Basic Notions of Dyadic Predicative Second-Order Logic

TERMINOLOGY	SYMBOL	DEFINITION
Empty Relation	E	$= «x, y: x \neq x \,\&\, y \neq y»$
Adjunction	$R^\frown\langle u, v\rangle$	$= «x, y: Rxy \vee (x = u \,\&\, y = v)»$
Deletion	$R \setminus \langle u, v\rangle$	$= «x, y: Rxy \,\&\, {\sim}(x = u \,\&\, y = v)»$
Identity	I_X	$= «x, y: Xx \,\&\, y = x»$
Distinctness	J_X	$= «x, y: Xx \,\&\, y \neq x»$
Image	$R[X]$	$= «y: \exists x(Xx \,\&\, Rxy)»$
Section	$R[x]$	$= «y: Rxy»$
Restriction	$R \mid X$	$= «x, y: Rxy \,\&\, Xx»$

TABLE J Modified Fregean Systems

THEORY	§ INTRODUCED	STRENGTH
PV ≈ ℘V	2.1, 2.4	$\geq I\Delta_0$, $< I\Delta_0$(superexp)
P^2V	2.3a	$\geq I\Delta_0$(exp), $< I\Delta_0$(superexp)
$P^\omega V$	2.3a	$\geq I\Delta_0$(exp), $< I\Delta_0$(super^{2}exp)
PHP	2.3c	$\geq I\Delta_0$, $< I\Delta_0$(superexp)
P^2HP	2.3c	$\geq I\Delta_0$(exp), $< I\Delta_0$(superexp)
$\wp^2V$	2.4	$\geq PA^2$
Π^1_1-FA	3.1	Π^1_1-CA_0
FA	3.1	PA^2
FA + *NV*	3.2	PA^3
$WFTA^2 \approx SFTA^2$	3.3–3.4	PA^3
Parsons set theory	3.5	PA^2
Boolos set theory	3.5	PA^2
ditto + ∞	3.5	PA^3
FB	3.6	ZF + indescribables

Notes

Chapter 1

Frege, Russell, and After

1. There is another usage of Frege's from which even the most faithful recent disciples of Frege have departed, and from which I have therefore felt free to depart myself, his use of "names of truth-values" for "sentences." This usage reflects a view that the category S is included in the category N, and hence that the truth-values are included among the objects, and the concepts among the functions. I have also ignored Frege's view that "identity" is literally applicable only to objects, which leads him to write of coextensiveness as the "analogue" of identity for concepts.
2. In Boolos, Burgess, and Jeffrey (2002), these topics are treated in chapter 9, which also contains further discussion of the topic of abbreviation, to be addressed shortly in the text.
3. In Boolos, Burgess, and Jeffrey (2002), this topic is treated in chapter 14. Here, however, our concern will be exclusively with axiomatizable theories.
4. Russell allows the ι-operator to be applied even where an existence and uniqueness presupposition does not hold, though in this case the equivalence of the existential with the universal unpackings of the definition, and of the result of first negating and then unpacking with the result of first unpacking and then negating, may not hold, and one does not in general have $\phi(\iota x\phi(x))$ or $\forall y\psi(y, \iota x\psi(y,x))$. I will never "give a name to" anything unless existence and uniqueness have been proved or assumed.

5. Or more precisely, an instance is a case of (5) where $\phi(x)$ and $\phi(y)$ are the results of substituting the variables x and y for the variable v in $\phi(v)$, where $\phi(v)$ is any formula. Or more precisely still, where $\phi(v)$ may be *almost* any formula, since we must exclude formulas like $\forall x(v = x)$ or $\exists y(v = y)$, where substituting x or y would result in that variable being "caught" by a quantifier. This sort of nonsense-excluding proviso will be left tacit in future formulations.
6. Or rather, of *pure* higher-order logic. In a non-logical theory based on a higher-order logic there might also be constants $a, b, c, \ldots$ for objects, and/or $A, B, C, \ldots$ for concepts, and/or $F, G, H, \ldots$ for relational concepts, and/or **A, B, C**,... for higher concepts, and so on, which could be substituted for variables of the appropriate kinds.
7. Also, to avoid nonsense, the variable X in (6) must not appear free in $\phi(x)$. This will be tacitly understood in all subsequent versions of comprehension.
8. Frege himself habitually avoided making existence assumptions explicit. He did not even have an existential quantifier, but always expressed $\exists$ as $\sim\forall\sim$. Thus he did not explicitly state comprehension in the form (6) above, but rather left it implicit in a rule permitting the substitution of a formula $\phi(x)$ for Xx or Yx or Zx. From this (6) follows, making such a substitution for Yx in the tautology $\sim\forall X\sim\forall x(Xx \leftrightarrow Yx)$. Recent writers have generally not followed Frege in this matter.
9. See Boolos (1985b).
10. In the decades immediately following Frege's work it was so cited especially by Henri Poincaré, one of the greatest mathematicians of the period. See Steiner (1975, chapter 1).
11. Actually, Frege generally states everything now conventionally stated in terms of successors in terms of predecessors instead. So he defines the less-than relation, which is the converse of the greater-than relation, as the ancestral of the predecessor relation, which is the converse of the successor relation. The whole preliminary account of Frege's procedure in the present chapter is only a rough approximation to the historical reality. Moreover, on the technical side I am slurring over some subtle but important points that will be gone into when we come back for a closer look in chapter 3.

12. For information on such operators see Corcoran et al. (1971, 1972). Frege's own procedure is not quite the same as any of the four I have described, but is closest to the fourth. He does, however, cling to the subordination principle insofar as he assumes not (10) but only the special case where $\phi(x)$ and $\psi(x)$ are of the forms Yx and Zx, the other cases following by his rule permitting the substitution of formulas for second-order variables.
13. If we want an equivalence on *all* objects, as Frege generally does, we can consider the relational concept under which fall any x and y such that either both are lines in the plane and they are parallel, or neither is a line in the plane. We will then have one extra abstract, which might be called *the undirected*, besides the directions.
14. See Dedekind (1888) and the commentary in Boolos (1994a).
15. Except in private correspondence. See Hallett (1984) for a thorough discussion.
16. This account of Russell's route to his paradox is given in Russell (1903); it is not mentioned in Russell (1902).
17. The proof just given is a free adaptation of Quine (1955), which was a refinement of Geach's reconstruction of an unpublished argument of Leśniewski.
18. For a fuller account, see Chihara (1973, chapter 1).
19. And no higher-level variables at all, a nonsense-excluding proviso that will be left tacit for the remainder of this exposition.
20. When one takes into consideration higher-level variables, and in particular when one takes into consideration that a first-level concept may be specified by a formula involving quantification over higher-level concepts, nothing so simple as numeral superscripts will suffice to distinguish all the types that need to be distinguished; but as already indicated, attention is being restricted to first-level concepts here.
21. For a sympathetic account of Russell's handling of infinity, see Boolos (1994b).
22. The key reason this is so is that classical mathematics can already prove the negation of any finitistically meaningful statement that is false, so that if it implied any finitistically meaningful statement that is false, it would imply a contradiction.
23. Unfortunately, Friedman's work remains unpublished. There is some information about it available in Harrington et al. (1985).

24. Actually, (Q1*) and (Q2*) render (Q1) redundant, but no use will be made of this fact. Q* can be interpreted in Q, by defining $x < y$ as $\exists z(z' + x = y)$. This definition permits the deduction of (Q1*)–(Q3*) as theorems. It is in this sense that Q and Q* are equivalent.
25. The general definition of interpretation on the one hand does not require (C<), but on the other hand requires something stronger than (R3*), namely the assumption that axiom (Q3*) holds with quantifiers relativized to δ, thus:

$$\delta(x) \rightarrow (x = 0 \vee \exists y(\delta(y) \,\&\, y < x \,\&\, x = y'))$$

But this relativized (Q3*) is something that follows immediately from (C<) and (R3*), which are far more convenient to work with in practice.
26. Actually, (Q3) is immediately provable once one has quantifier-free induction, and may be dropped from the list of axioms as redundant.
27. Q_ω with symbols for all primitive recursive functions introduced is an extension of another system, one that is called *Skolem arithmetic* or PRA for *primitive recursive arithmetic*, and that is generally accepted as formalizing what is *directly* acceptable from a finitist standpoint; but this latter system is not a first-order theory, and for that reason will be left out of account here.
28. More specifically, if symbols for primitive recursive functions are appropriately added to $I\Sigma_1$, we get an extension of Q_ω that is conservative for Π_2 sentences over it, and for that matter over the system PRA that I have omitted from my discussion. This conservativeness result was obtained not only by Parsons, but also independently—though the dates of publication are later—by Grigori Mints and by Gaisi Takeuti, all three using different methods.
29. In Boolos, Burgess, and Jeffrey (2002), the Ackermann example is treated at the end of chapter 7.
30. Except that for technical reasons the weakest of them, which will be of little concern here, assumes Σ_1^0-induction as in the system $I\Sigma_1$ of arithmetic.
31. The weakest of them is precisely the system PRA, which formalizes *finitism*. The philosophical 'isms corresponding to the F_i for $i > 1$ are *intuitionism minus choice sequences*, *predicativism*, and

predicativism plus inductive definitions, as formalized in systems called HA, IR, and $ID_{<\omega}$, but these will not be discussed here.

32. The history of this result is complicated. The first writers on the topic were solely or mainly concerned with proving the set theory NBG conservative over the set theory ZFC, though the result is quite general and applies, for instance, to show that the fragment of analysis $F_2 = \Sigma^0_1\text{-}CA_0 = \Delta^1_0\text{-}CA_0 = ACA_0$ is conservative over the arithmetic $P^1 = PA$. Model-theoretic proofs were offered by Mostowski (1950), Novak Gal (1951), and Rosser and Wang (1950). Shoenfield (1954) gave a proof-theoretic proof, and I have applied his name to the theorem because it is his version that I am going to be using in the next chapter. In Boolos, Burgess, and Jeffrey (2002) the model-theoretic proof is given for the analysis/arithmetic case in chapter 25, section 3.
33. In the case of the Reinhardt cardinal, one of the new theorems is $0 = 1$. In the case of any of the other systems, the statement of the consistency of any of the systems weaker than itself is, or becomes when statements about formulas are coded as statements about numbers, a Π^0_1 formula implied by that system and not the weaker one. However, coded consistency statements, though the easiest example to cite when addressing a non-specialist audience, do not by any means exhaust the "down-to-earth" consequences of large cardinals.
34. To name names, Crisipin Wright, Bob Hale, and their associates are neo-Fregeans in the narrow sense, while George Boolos, Richard Heck, Kit Fine, and many others are neo-Fregeans in the broad sense (only). Some may prefer "thin" *vs* "thick" to my "broad" *vs* "narrow."
35. Settling the place of a theory on the scale from Robinson to Reinhardt means settling its *proof-theoretic* status, but there do remain for some of the approaches other logical questions, about *model-theoretic* features, that will not be addressed in the present study.
36. From which it is concluded at the end of Bays (2000) that we "need to spend a lot more time doing neo-logicist mathematics . . . before we can productively evaluate the philosophical significance of this mathematics," a sentiment with which I heartily agree.
37. I write "notion" or "idea" where it might be more natural to write "concept" in order to avoid confusion with "concepts" in Frege's

technical sense of that term, whose very existence is, of course, one of the issues under debate. I have even avoided the term "conceptual truth," which many recent writers prefer, in favor of the old-fashioned "analytic."

38. The same is true for the assumption of the existence of abstracts for *any* equivalence *on objects*, provided the abstracts are not themselves taken to belong to the original domain of objects. The proof will be indicated late in the next chapter.
39. In the case of the most philosophically ambitious neo-Fregean school, the issue of what is special about *its* preferred strategy will be remarked upon and at somewhat greater length than most other philosophical issues at the appropriate place in the survey to follow, under the label "the *bad company* objection" (section 3.2).
40. At least in application to neo-Fregeanism. For it parallels a terminology introduced long ago in connection with the issue of nominalism. Probably "reformist" would be better than "revolutionary" in the present context, but I will keep "revolutionary" for the sake of the parallel.
41. Whether or not this was Cantor's view, a hybrid cardinal-ordinal view of numbers has been defended on empirical grounds by the famous developmental psychologist Jean Piaget in his contribution to Beth and Piaget (1966). Of course, if one is engaged as Frege was in a project of "proving whatever can be proved," it is hard to see how else one could proceed, in the case of a hybrid or "cluster" concept, *except* by selecting and "privileging" *one* of its aspects as a definition, and deriving the others from it.
42. He annoyed Cantor by objecting in §86 of the *Grundlagen* to Cantor's use of the word "number" in connection with his ordinals. A disproportionate amount of space is devoted to this issue in the short review Cantor (1885), where Cantor claims both that ordinary usage does not support Frege's position as Frege had claimed, and that it would not matter if it did. It may well be that irritation over this point was at least partly responsible for the unfortunate tone of the review as a whole.
43. It is all very well to hold, as Frege did and many recent writers such as Wright (2000) do, that an account that builds the applicability of the notion of natural number into its very definition is better than one that does not; and it may well be that abstractionist

definitions do better than set-theoretic definitions in this regard; but the cardinalist definition does not obviously do better than the ordinalist, or vice versa. Moreover, even if one adheres to a cardinalist approach, this itself comes, as Marco Lopez reminded me, in different versions in the *Grundlagen*, which does not exploit the possibility of type-lowering, and *Grundgesetze*, which does.

44. More generally, see Russell's remarks towards the end of section II, "The Definition of Number," preceding his displayed definition of the number of a class. This material is most readily available in the reprint of selections in Benacerraf and Putnam (1983), with the quoted passage appearing on page 172.
45. As for leaving something out, the view that *counting* is at least *a* crucial ingredient of the number-concept has been maintained by philosophers from Husserl to Heck (2000), whose discussion of this issue I recommend to the reader. As for putting too much in, that is the focus of Benacerraf's criticism of set-theoretic definitions, and as for Frege's definition, even so sympathetic a commentator as Dummett admits that "any claim to have captured the meaning attached to phrases of the form 'the number of *F*s' by ordinary speakers of the language is palpably absurd." Yet Dummett tries to explain how Frege could, despite this fact, nonetheless consider himself to be proving *the same laws* that earlier mathematicians had accepted without proof. (I confess to having not fully followed the explanation. See Dummett (1991), last section of chapter 14, where the quoted passage appears on page 177.)

Chapter 2

Predicative Theories

1. Certain remarks of Poincaré were also influential, but must be left out of account here. See Chihara (1973, chapter 4).
2. Burgess and Hazen (1998). Hazen has besides been throughout my main source for historical information and references to the earlier literature.
3. In Boolos, Burgess, and Jeffrey (2002), the theorem is covered in chapter 21, sections 1 and 2.

4. For 0 bound variables, all there will be to the conjunction will be statements of the first and second kind. For 0 free variables, all there will be to the conjunction will be statements of the third kind to the effect that there are exactly k objects or at least n objects v for which the combination holds. For 0 one-place relation symbols, all there will be will be statements to the effect that there are exactly k or at least n objects, *tout court*.
5. The statement that there are exactly k or at least n objects, *tout court*, may be taken to be $\exists_k!v(v = v)$ or $\exists_n v(v = v)$. These are not the shortest formulas expressing what is wanted, but will do.
6. The oldest published source known to me is Quine (1961).
7. Notably in work of Feferman and Hellman (1995) and Ferreira (200?). A different sort of approach that also leads to first-order Peano arithmetic can be found in Antonelli and May (200?).
8. See Urquhart (1994).
9. A proof will be given for a very similar result in the first section of the next chapter.
10. It will be seen in the next section that there are also *non*-standard models of full comprehension. In the literature the term *Henkin model*, avoided here, is used sometimes for all models of full comprehension, and sometimes just for the non-standard ones.
11. In particular, in Wang (1952) he noted that for T = ST, second-order Peano arithmetic P^2 can be interpreted in T', and vice versa. As I stated earlier, this is actually true for UST as well.
12. It may mentioned without proof here that though $\wp$V is ostensibly stronger than PV, it is not really so. For with just = in the original first-order language, and with only monadic concept variables, it can be shown as an extension of the Löwenheim-Behmann theorem that any formula *with* bound monadic concept variables is provably equivalent to one with*out* bound concept variables. It may be added, however, that the theory $\wp^2$V I am about to introduce is not just ostensibly but really considerably stronger than P^2V. Indeed, it follows easily from Dedekind's Theorem, Version II, that it can interpret P^2. Proving consistency for $\wp^2$V will suffice to prove consistency for P^2V, but it will in a sense be overkill.
13. The argument to this point is an adaptation of the well-known "Skolem paradox." Compare Boolos, Burgess, and Jeffrey (2002, pp. 252–53).

14. In the case of the Löwenheim-Behmann theorem, the original proof was comparatively elementary and formalizable in Q_3. Shoenfield used a result related to Gentzen's theorem, which requires Q_4. The original proof in Craig (1957) used Gentzen's theorem itself. Boolos, Burgess, and Jeffrey (2002) gives a model-theoretic treatment in chapter 20 that derives from H. J. Keisler.
15. Actually, for the very simple language L_0, the consistency of T_1 could be proved more directly than by appeal to Shoenfield's theorem, but the proof I am giving is one that will work with other starting theories than T_0.
16. Though it is also true that so far no one has found a way to exploit the extra complication and interpret more than Q_3 in them, either.
17. Note that there is no *general* way to define elementhood in T_0, and there had better not be, because if there were, we would have to confront $\{x: x \notin x\}$. The definition in (3) works only because we are restricting attention to those y for which (2) holds. Goldfarb (2001) proves $T_0{}^*$ undecidable, another indication of some degree of latent strength.
18. This seems to hold for the usual proof-theoretic proof as well, which would make possible a finitistic proof of Heck's consistency result, but so far no one has published a fully worked-out treatment. Note that usually the conclusion of Shoenfield's theorem is formulated as stating that the axiom scheme (5) may be replaced by the single axiom

$$(\forall X)\,(\forall Y)\text{—}Xx\text{—}Yx\text{—}$$

but from this axiom and (6) there of course follow all instances of (5) for ϕ and ψ without bound concept variables.
19. Unfortunately Wehmeier's method of proof, based on an extension of the Löwenheim-Behmann theorem, does not seem to extend to P^2V with Δ^1_1-comprehension.
20. Their method of proof seems to extend to $T_n{}^*$ for $n > 1$, though they do not explicitly discuss the matter.
21. For the cognoscenti, they use the Barwise-Schlipf apparatus of recursively saturated models. However, in other contexts there is a proof-theoretic substitute even for applications of this apparatus, but so far no one has published a fully worked-out proof of such a substitute in the Fregean context.

Chapter 3

Impredicative Theories

1. Boolos (1987) was concerned with Frege's informal outline in the *Grundlagen*, Heck (1993) with the formal development in the *Grundgesetze*, Boolos and Heck (1998) with the differences between the two.
2. See Tennant (1987), Boolos (1996, note 9), Heck (1997a, 1997b), Bell (1999). There are two ways a finitude restriction might be implemented, either by making finitude of X a condition for the existence of an abstract for X with respect to equinumerosity, or by considering abstracts with respect to a coarser equivalence agreeing with equinumerosity for finite X, but counting all infinite X as equivalent. The latter implementation differs from the former only in supplying an object, which may be called *infinity* ∞, as the abstract common to all infinite X.
3. It may be mentioned that the predicative analogue of Π^1_1-P^2, which is to say dyadic predicative second-order logic with (Q1) and (Q2), is *not* strong enough to interpret the predicative analogue of Π^1_1-CA_0, which is to say Δ^1_0-CA_0 (the Friedman system F_2). The latter system is more than strong enough to prove the consistency of the former.
4. As a hint, the first step is to show for any X by induction on z that there exists a number coding a sequence that lists the x in the intersection of X and $\langle z \rangle$ in increasing order. (Note that the hypothesis on which one is performing induction involves X as a parameter but involves no bound second-order variables.) If X is bounded, which is to say, contained in some $\langle z \rangle$, we then get a sequence coding a one-to-one correspondence between X and some $\langle y \rangle$ with $y \leq z$; whereas if X is unbounded, it can be shown that the various sequences can be fitted together to give a one-to-one correspondence between U and X. (Recalling that we have already remarked that the reflexivity, symmetry, and transitivity of equinumerosity can be proved using only predicative comprehension, this shows that all unbounded X are equinumerous.) Various uniqueness claims then have to be established to complete the proof.
5. That a non-trivial amount of number theory can in a sense be done even without the existence of numbers is emphasized in a

series of papers by Andrew Boucher that have as of the time of this writing not yet been published, and that came to my attention too late to be made use of here. Preprints are available at www.andrewboucher.com.

6. That we have such "expressive conservativeness" as well as "deductive conservativeness" is further explained in §I.B.2.b of Burgess and Rosen (1997).
7. This question has been endlessly discussed. See, for instance, chapter 14, "To Bury Caesar . . . " in Wright and Hale (2001); §II.3, "The Caesar Problem" in Fine (2002); and Heck (1997c). Heck's proposal, the details of which there is not space present here, actually results in a variant of Frege's that is deductively conservative and in which the Julius Caesar question is ill-formed.
8. With a further special assumption, a sufficiently stong form of the axiom of choice, this question is equivalent to the question whether there exists a relational concept *R* such that for no two distinct objects *u* and *v* are «*x*: *Rux*» and «*x*: *Rvx*» equinumerous. But without special assumptions, the question is not provably equivalent to any question not about numbers. Even the proposal of Heck (1997c) leaves us with this counterexample to expressive conservativeness.
9. This makes (2) a *sufficient* condition for non-inflating. The *natural* formulation of non-inflating would require *third*-order logic; and the proof that (2) is a *necessary* condition for non-inflating as naturally defined would require a version of the axiom of choice. In the literature non-inflation is sometimes called "conservativeness," a deplorable misusage of that technical term from logic.
10. The idea goes back to Mostowski (1957), and has become well known to philosophers since the postumous publication of Tarski (1986). For a recent discussion with extensive further references, see Feferman (1999).
11. Officially, languages are not supposed to have rules permitting the introduction of new symbols, but are supposed to have all their symbols given in advance. In case the reader is troubled by this point, a language of the type required can be constructed as follows. First enumerate all the formulas we have before introducing any special symbols as θ_1, θ_3, $\theta_5, \ldots$, then introduce $\S_1$ associated with θ_1, then enumerate all the new formulas we get with this

symbol as $\theta_2, \theta_6, \theta_{10}, \cdots$, then introduce $\S_2$ associated with θ_2, then enumerate the new formulas we get with this symbol as θ_4, θ_{12}, $\theta_{20}, \cdots$, and continue in this way, each time using up half the remaining places. The result is a language with a symbol $\S_i$ for every formula θ_i, and with each θ_i containing only $\S_j$ for $j < i$.

12. This characterization should be qualified by a recognition that Fine's approach does indirectly make surrogates for certain extensions available, as we have seen. The "Parsons set theory" to be introduced shortly is directly related to the general theory of abstraction as well.
13. It is a well-known, though not well-understood, phenomenon of mathematical psychology that often the mere announcement that a genius has proved some result, without any indication of the proof, may suffice (a) to inspire a sub-genius to produce a proof of the same result; and (b) to inspire a non-genius to produce a proof of a partial result with a stronger hypothesis and/or weaker conclusion. The present material represents an instance of phenomenon (b). The result expounded in this section was as much as I was able to work out for myself on learning of a stronger result of Harvey Friedman, without actually seeing his proofs—and without at the time being aware of the earlier work of Bernays, in whose footsteps I had been unwittingly following all the while. I am grateful to Kai Wehmeier for bringing Bernays's paper to my attention, to Stephan Leuenberger for help while I was struggling with the German, and to Solomon Feferman for pointing out that the struggle was unnecessary, since there exists an English translation. Another work of which I became aware rather late is Shapiro (2003), which ends with an allusion to Bernays's approach and its possible relationship to neo-Fregean projects.
14. See Hrbacek and Jech (1999), chapter 15.
15. Really in (6) R ought to be restricted to relations such that Ruv implies Xu and Xv for all u and v, but this technical point is irrelevant to the main conceptual issue, and will be ignored below.
16. Officially we are still working with monadic second-order logic, so the dyadic second-order quantifier is really an abbreviation for something involving a monadic second-order quantifier and mention of ordered pairs, but this technical point is irrelevant to the main conceptual issue, and will be ignored below.

17. It may be mentioned parenthetically that *many but not all* new set-theoretic axioms would be equivalent to new logical axioms. The truth of any conjecture in number theory expressible as a formula ψ of first- or second-order Peano arithmetic—the arithmetical formula coding the statement that FB is consistent, for instance—is equivalent to the standard and hence intuitive validity of a certain second-order formula, namely, $\phi \to \psi$ where ϕ is the conjunction of all the non-logical axioms of P^2, of which there are only finitely many, counting the comprehension axioms as logical rather than non-logical. The truth of the continuum hypothesis is equivalent to the validity of another second-order formula. But the truth of some large cardinal hypotheses is not.
18. For a different perspective on plural quantification in relation to modified Fregean systems, compare Shapiro and Weir (2000). For a neo-Fregean position that sees a genuine problem in the awkwardness for which Frege begs indulgence, but nonetheless does not see any serious problem about second-order quantification, compare Wright (1999).
19. For a "translation" of the material of the last several sections into the language of plural quantification, see Burgess (2004).

References

Ajdukiewicz, Kasimierz. 1935. "Die syntaktische Konnexität." *Studia Philosophica* 1:1–27.

Antonelli, Aldo, and Robert May. 200?. "Frege's Other Program." *Notre Dame Journal of Formal Logic*. Forthcoming.

Bar-Hillel, Yehoshua. 1950. "On Syntactical Categories." *Journal of Symbolic Logic* 15:1–16.

Bays, T. 2000. "The Fruits of Logicism." *Notre Dame Journal of Formal Logic* 41:415–21.

Behmann, Heinrich. 1922. "Beiträge zur Algebra der Logik, insbesondere zum Entscheidungsproblem." *Mathematische Annalen* 86:419–32.

Bell, J. L. 1999. "Frege's Theorem in a Constructive Setting." *Journal of Symbol Logic* 64:486–88.

Benacerraf, Paul. 1965. "What Numbers Could Not Be." *Philosophical Review* 74:47–73. Reprinted in Benacerraf and Putnam (1983), 272–94.

——.1973. "Mathematical Truth." *Journal of Philosophy*, 70:661–80. Reprinted in Benacerraf and Putnam (1983), 403–20.

Benacerraf, Paul, and Hilary, Putnam. 1983. *Philosophy of Mathematics: Selected Readings*, 2nd ed. Cambridge: Cambridge University Press.

Bernays, Paul. 1961. "Zur Frage der Unendlichkeitsschemata in der axiomatischen Mengenlehre." In *Essays on the Foundations of Mathematics*, ed. Yehoshua Bar-Hillel et al., 3–49. Jerusalem: Magnes Press. Trans. J. Bell and M. Planitz (1976) as "On the

Problem of Schemata of Infinity in Axiomatic Set Theory." In *Sets and Classes: On the Work by Paul Bernays*, ed. Gert Müller, 121–72. Amsterdam: North Holland.

Beth, E. W. and J. Piaget. 1966. *Mathematical Epistemology and Psychology*. Dordrecht: Reidel.

Boolos, George S. 1984. "To Be Is to Be the Value of a Variable (or Some Values of Some Variables)." *Journal of Philosophy* 81:430–50. Reprinted in Boolos (1998), 54–72.

——. 1985a. "Nominalist Platonism." *Philosophical Review* 94:327–44. Reprinted in Boolos (1998), 73–87.

——. 1985b. "Reading the *Begriffsschrift*." *Mind* 94:331–44. Reprinted in Demopoulos (1995), 163–81. Reprinted in Boolos (1999), 155–70.

——. 1987. "The Consistency of Frege's *Foundations of Arithmetic*." In *On Being and Saying: Essays for Richard Cartwright*, ed. J. J. Thomson, 3–20. Cambridge, MA: MIT Press. Reprinted in Demopoulos (1995), 211–33. Reprinted in Boolos (1998), 182–201.

——. 1989. "Iteration Again." *Philosophical Topics* 17:5–21. Reprinted in Boolos (1998), 88–204.

——. 1990. "The Standard of Equality of Numbers." In *Meaning and Method: Essays in Honor of Hilary Putnam*, ed. George Boolos, 261–78. Cambridge: Cambridge University Press. Reprinted in Demopoulos (1995), 234–54. Reprinted in Boolos (1998), 202–19.

——. 1993. "Whence the Contradiction?" *Aristotelian Society Supplement* 67:213–33. Reprinted in Boolos (1998), 220–36.

——. 1994a. "1879?" In *Reading Putnam*, ed. Peter Clark and Bob Hale, 31–48. Oxford: Blackwell. Reprinted in Boolos (1998), 237–54.

——. 1994b. "The Advantage of Honest Toil over Theft." In *Mathematics and Mind*, ed. Alexander George, 27–44. Oxford: Oxford University Press. Reprinted in Boolos (1998), 255–74.

——. 1996. "On the Proof of Frege's Theorem." In *Benacerraf and His Critics*, ed. Adam Morton, 143–59. Oxford: Blackwell. Reprinted in Boolos (1998), 275–90.

——. 1998. *Logic, Logic, and Logic*. Cambridge, MA: Harvard University Press.

Boolos, George S., John P. Burgess, and Richard C. Jeffrey. 2002. *Computability and Logic*, 4th edition. Cambridge: Cambridge University Press.

Boolos, George S., and Richard, Heck. 1998. "*Die Grundlagen der Arithmetik* §§82–3." In *Philosophy of Mathematics Today*, ed. Matthias Schirn, 407–28. Oxford: Clarendon Press. Reprinted in Boolos (1998), 313–38.

Burali-Forti, Cesare. 1967 (originally published 1897). "A Question of Transfinite Numbers / On Well-Ordered Classes." Trans. J. van Heijenoort. In van Heijenoort (1967), 105–12.

Burgess, John P. 1984. Review of Wright (1983). *Philosophical Review* 93:638–40.

——. 2004. "E *Pluribus Unum*." *Philosophia Mathematica* 12: 193–221.

Burgess, John P., and A. P. Hazen, 1998. "Arithmetic and Predicative Logic." *Notre Dame Journal of Formal Logic* 39:1–17.

Burgess, John P., and Gideon, Rosen. 1997. *A Subject with No Object: Strategies for Nominalistic Interpretation of Mathematics*. Oxford: Oxford University Press.

Cantor, Georg. 1885. Review of Frege (1884). *Deutsche Literaturzeitung* 6:728–29.

Chihara, Charles. 1973. *Ontology and the Vicious Circle Principle*. Ithaca, NY: Cornell University Press.

Collins, George F., and J. D. Halpern. 1970. "On the Interpretability of Arithmetic in Set Theory." *Notre Dame Journal of Formal Logic* 11:477–83.

Corcoran, John, and J. Herring. 1971. "Notes on a Semantic Analysis of Variable-Binding Term Operators." *Logique et Analyse* 55: 664–57.

Corcoran, John, W. Hachter, and J. Herring, 1972. "Variable-Binding Term Operators." *Zeitschrift für mathematische Logik und Grundlagen der Mathematik* 18:177–82.

Craig, William. 1957. "Three Uses of the Herbrand-Gentzen Theorem in Relating Model Theory and Proof Theory." *Journal of Symbolic Logic* 22:269–85.

Dedekind, Richard. 1888. *Was sind und was sollen die Zahlen?* Braunschweig: Vieweg. Trans. Wooster Woodruff Beman (1901) as "The Nature and Meaning of Numbers." In *Essays on the Theory of Numbers*, 29–115. Chicago: Open Court.

Demopoulos, William, ed. 1995. *Frege's Philosophy of Mathematics*. Cambridge, MA: Harvard University Press.

Dummett, Michael. 1991. *Frege: Philosophy of Mathematics*. Cambridge, MA: Harvard University Press.

Feferman, Solomon. 1999. "Logic, Logics, and Logicism." *Notre Dame Journal of Formal Logic* 40:31–54.

Feferman, Solomon, and Geoffrey, Hellman. 1995. "Predicative Foundations of Arithmetic." *Journal of Philosophical Logic* 24:1–17.

Ferreira, Fernando. 200?. "Amending Frege's *Grundgesetze der Mathematik*." *Synthese*. Forthcoming.

Ferreira, Fernando, and Kai Wehmeier. 2002. "On the Consistency of the Δ^1_1-CA Fragment of Frege's *Grundgestze*." *Journal of Philosophical Logic* 31:301–11.

Fine, Kit. 2002. *The Limits of Abstraction*. Oxford: Oxford University Press.

Frege, Gottlob. 1879. *Begriffsschrift: Eine der arithmetischen nachgebildete Formelsprache des reinen Denkens*. Halle: Louis Nebert. Trans. Stefan Bauer-Mengelberg. In van Heijenoort (1967), 1–82.

——. 1884. *Die Grundlagen der Arithmetik: Eine logisch-mathematische Untersuchung über den Begriff der Zahl*. Breslau: Wilhelm Koebner. Trans. J. L. Austin. *Foundations of Arithmetic*. 1950. Oxford: Blackwell.

——. 1892. "Über Sinn und Bedeutung." *Zeitschrift für Philosophie und philosophische Kritik*. 100:25–50. Trans. Max Black as "On Sense and Reference." In Geach and Black (1960), 56–78.

——. 1893/1903. *Grundgesetze der Arithmetik, begriffsschriftlich abgeleitet*. 2 vols. Jena: Pohle. Reprinted 1962. Hildesheim: Olms. Partial trans. P.E.B. Jourdain and J. Stachelroth as *Basic Laws of Arithmetic*. In Geach and Black (1960), 137–244.

Geach, Peter. 1975. Review of M. Dummett, *Frege: Philosophy of Language*. *Mind* 84:436–99.

Geach, Peter, and Max Black, eds. 1960. *Translations from the Writings of Gottlob Frege*. Oxford: Blackwell.

Goldfarb, Warren. 2001. "First-Order Freagean Theory is Undecidable." *Journal of Philosophical Logic* 30:613–16.

Hájek, Petr, and Pavel Pudlak. 1998. *Metamathematics of First-Order Arithmetic*. Berlin: Springer.

Hale, Bob. 2000. "Reals by Abstraction." *Philosophia Mathematica* 8:100–123.

Hallett, Michael. 1984. *Cantorian Set Theory and Limitation of Size.* Oxford: Oxford University Press.

Harrington, Leo, et al., eds. 1985. *Harvey Friedman's Research on the Foundations of Mathematics.* Amsterdam: North Holland.

Hazen, A. P. 1985. Review of Wright (1983). *Australasian Journal of Philosophy* 63:251–54.

Heck, Jr., Richard G. 1992. "On the Consistency of Second-Order Contextual Definitions." *Noûs* 26:491–95.

——. 1993. "The Development of Arithmetic in Frege's *Grundgesetze der Arithmetik.*" *Journal of Symbolic Logic* 58:579–601. Reprinted with a postscript in Demopoulos (1995), 257–94.

——. 1996. "On the Consistency of Predicative Fragments of Frege's *Grundgesetze der Arithmetik.*" *History and Philosophy of Logic* 17:209–20.

——. 1997a. "The Finite and the Infinite in Frege." In *Philosophy of Mathematics Today*, ed. M. Schirn, 429–66. Oxford: Oxford University Press.

——. 1997b. "Finitude and Hume's Principle." *Journal of Philosophical Logic* 26:589–617.

——. 1997c. "The Julius Caesar Objection." In *Language, Thought, and Logic: Essays in Honour of Michael Dummett*, ed. R. Heck. 273–308. Oxford: Oxford University Press.

——. 2000. "Cardinality, Counting, and Equinumerosity." *Notre Dame Journal of Formal Logic* 41:87–109.

Hodes, Harold. 1984. "Logicism and the Ontological Commitments of Arithmetic." *Journal of Philosophy* 81:123–49.

Hrbacek, Karel, and Thomas Jech. 1999. *Introduction to Set Theory*, 3rd ed. New York: Marcel Dekker.

Kanamori, Akihiro. 1997. *The Higher Infinite*, corrected printing. Berlin: Springer.

Linnebo, Øystein. 2004. "Predicative Fragments of Frege Arithmetic." *Bulletin of Symbolic Logic* 10:153–74.

Löwenheim, Leopold. 1915. "Über Möglichkeiten im Relativkalkül." *Mathematische Annalen* 76:447–70. Trans. Stefan Bauer-Mengelberg as "On Possibilities in the Calculus of Relatives." In van Heijenoort (1967), 228–51.

Montagna, Franco, and Antonella, Mancini. 1994. "A Minimal Predicative Set Theory." *Notre Dame Journal of Formal Logic* 35:186–203.

Mostowski, Andrzej. 1950. "Some Impredicative Definitions in the Axiomatic Set Theory." *Fundamenta Mathematicæ* 36:143–64.

——. 1957. "On a Generalization of Quantifiers." *Fundamenta Mathematicæ* 44:12–36.

Nelson, Edward. 1986. *Predicative Arithmetic*. Princeton: Princeton University Press.

Novak Gal, Ilse. 1951. "A Construction of Models of Consistent Systems." *Fundamenta Mathematicæ* 37:87–110.

Parsons, Charles. 1965. "Frege's Theory of Number." In *Philosophy in America*, ed. Max Black, 180–203. Ithaca, NY: Cornell University Press. Reprinted in Demopoulos (1995), 182–207.

Parsons, Terence. 1987. "On the Consistency of the First-Order Portion of Frege's Logical System." *Notre Dame Journal of Formal Logic* 28:61–68. Reprinted in Demopoulos (1995), 422–31.

Quine, W. V. 1951. *Mathematical Logic*, revised ed. Cambridge, MA: Harvard University Press.

——. 1955. "On Frege's Way Out." *Mind* 64:145–59. Reprinted in *Selected Logic Papers*. 1966. 146–58. New York: Random House.

——. 1961. "A Basis for Number Theory in Finite Classes." *Bulletin of the American Mathematical Society* 67:391–92.

Ramsey, F. P. 1925. "The Foundations of Mathematics." *Proceedings of the London Mathematical Society* 25:338–84. Reprinted in *The Foundations of Mathematics* (1931), ed. R. B. Braithwaite, 1–61. London: Paul, Trench, Tubner.

Rayo, Augustín, and Gabriel Uzquiano. 1999. "Toward a Theory of Second-Order Consequence." *Notre Dame Journal of Formal Logic* 40:315–25.

Rosser, J. Barkley. 1942. "The Burali-Forti Paradox." *Journal of Symbolic Logic* 7:1–17.

Rosser, J. Barkley, and Hao Wang. 1950. "Non-Standard Models for Formal Logics." *Journal of Symbolic Logic* 16:113–29.

Russell, Bertrand. 1902. Letter to Frege, 16 June. Trans. Beverly Woodward. In van Heijenoort (1967). 124–25.

——. 1903. *The Principles of Mathematics* I. Cambridge: Cambridge University Press.

——. 1906. "Some Difficulties in the Theory of Transfinite Numbers and Order Types," *Proceedings of the London Mathematical Society* 2:29–53.

——. 1908. "Mathematical Logic as Based on the Theory of Types." *American Journal of Mathematics* 30:222–62. Reprinted in van Heijenoort (1967), 150–82.

Schroeder-Heister, Peter. 1987. "A Model-Theoretic Reconstruction of Frege's Permutation Argument." *Notre Dame Journal of Formal Logic* 28:69–79.

Shapiro, Stewart. 1987. "Principles of Reflection and Second-Order Logic." *Journal of Philosophical Logic* 16:309–33.

——. 2000. "Frege meets Dedekind: A Neo-Logicist Treatment of Real Analysis." *Notre Dame Journal of Formal Logic* 41:335–64.

——. 2003. "Prolegomena to any Future Neo-Logicist Set Theory: Abstraction and Indefinite Extensibility." *British Journal for the Philosophy of Science* 54:59–91.

Shapiro, Stewart, and Alan Weir. 1999. "New V, ZF, and Abstraction." *Philosophia Mathematica* 7:293–321.

——. 2000. "Neo-Logicist Logic Is Not Epistemically Innocent." *Philosophia Mathematica* 8:160–89.

Shoenfield, J. R. 1954. "A Relative Consistency Proof." *Journal of Symbolic Logic* 19:21–28.

Simpson, Stephen G. 1999. *Subsystems of Second Order Arithmetic*. Berlin: Springer.

Steiner, Mark. 1975. *Mathematical Knowledge*. Ithaca, NY: Cornell University Press.

Tarski, Alfred. 1986. "What Are Logical Notions?" *History and Philosophy of Logic* 7:143–54.

Tarski, Alfred, and Andrzej Mostowski, and Raphael M. Robinson. 1953. *Undecidable Theories*. Amsterdam: North Holland.

Tennant, Neil. 1987. *Anti-Realism and Logic: Truth as Eternal*. Oxford: Oxford University Press.

Urquhart, Alasdair. 1994. Editorial introductions to *Collected Papers of Bertrand Russell*, 4: *Foundations of Logic 1903–05*. London: Routledge.

Van Heijenoort, Jean, ed. 1967. *From Frege to Gödel: A Sourcebook in Mathematical Logic, 1879–1931*. Cambridge, MA: Harvard University Press.

Wang, Hao. 1952. "Truth Definitions and Consistency Proofs." *Transactions of the American Mathematical Society* 73:243–75.

Wehmeier, Kai. 1999. "Consistent Fragments of the *Grundgesetze* and the Existence of Non-Logical Objects." *Synthese* 121:309–28.

Whitehead, Alfred North, and Bertrand Russell. 1910–13. *Principia Mathematica*, 3 vols. Cambridge: Cambridge University Press. 2nd ed. 1925–27.

Wright, Crispin. 1983. *Frege's Conception of Numbers as Objects*. Aberdeen: Scots Philosophical Monographs.

——. 1999. "Why Frege Does Not Deserve His Grain of Salt: A Note on the Paradox of 'The Concept Horse' and the Ascription of *Bedeutungen* to Predicates." In *New Essays on the Philosophy of Michael Dummett* (Gratzer Philosophische Studien 55), eds. J. Brandl and P. Sullivan, 239–63. Vienna: Rodopi.

——. 2000. "Neo-Fregean Foundations for Real Analysis: Some Reflections of Frege's Constraint." *Notre Dame Journal of Formal Logic* 41: 317–34.

Wright, Crispin, and Bob Hale. 2001. *The Reason's Proper Study: Essays towards a Neo-Fregean Philosophy of Mathematics*. Oxford: Oxford University Press.

Index